Lecture Notes in Computer Science   13767

Founding Editors

Gerhard Goos

Juris Hartmanis

The series Lecture Notes in Computer Science (LNCS), including its subseries Lecture Notes in Artificial Intelligence (LNAI) and Lecture Notes in Bioinformatics (LNBI), has established itself as a medium for the publication of new developments in computer science and information technology research, teaching, and education.

LNCS enjoys close cooperation with the computer science R & D community, the series counts many renowned academics among its volume editors and paper authors, and collaborates with prestigious societies. Its mission is to serve this international community by providing an invaluable service, mainly focused on the publication of conference and workshop proceedings and postproceedings. LNCS commenced publication in 1973.

Éric Renault · Paul Mühlethaler

Editors

Machine Learning for Networking

5th International Conference, MLN 2022
Paris, France, November 28–30, 2022
Revised Selected Papers

 Springer

Editors
Éric Renault
ESIEE Paris
Noisy-le-Grand, France

Paul Mühlethaler
Inria
Paris, France

ISSN 0302-9743   ISSN 1611-3349 (electronic)
Lecture Notes in Computer Science
ISBN 978-3-031-36182-1   ISBN 978-3-031-36183-8 (eBook)
https://doi.org/10.1007/978-3-031-36183-8

This Springer imprint is published by the registered company Springer Nature Switzerland AG
The registered company address is: Gewerbestrasse 11, 6330 Cham, Switzerland

Preface

The rapid development of new network infrastructures and services has led to the generation of huge amounts of data, and machine learning now appears to be the best solution to process these data and make the right decisions for network management. The International Conference on Machine Learning for Networking (MLN) aimed at providing a top forum for researchers and practitioners to present and discuss new trends in deep and reinforcement learning, pattern recognition and classification for networks, machine learning for network slicing optimization, 5G systems, user behavior prediction, multimedia, IoT, security and protection, optimization and new innovative machine learning methods, performance analysis of machine learning algorithms, experimental evaluations of machine learning, data mining in heterogeneous networks, distributed and decentralized machine learning algorithms, intelligent cloud-support communications, resource allocation, energy-aware communications, software-defined networks, cooperative networks, positioning and navigation systems, wireless communications, wireless sensor networks, and underwater sensor networks.

The call for papers resulted in a total of 27 submissions from all around the world: Algeria, Austria, Canada, China, Ecuador, France, Morocco, Philippines, Saudi Arabia, Spain, Tunisia, and the USA. All submissions were assigned to at least five members of the Program Committee for double-blind review. The Program Committee decided to accept 12 papers.

The paper *Low Complexity Adaptive ML Approaches for End-to-End Latency Prediction* by Pierre Larrenie (Thales SIX and ESIEE Paris, France), Jean-François Bercher (ESIEE Paris, France), Iyad Lahsen-Cherif (INPT, Morocco) and Olivier Venard (Univ. Gustave Eiffel, France), was awarded the prize for the best paper.

Three keynotes and a tutorial completed the program: *Interpretable & Explainable Machine Learning (IML/XAI) in the Industrial IoT domain: Are Bayesian Optimization and IML/XAI far from each other?* by Soumya Banerjee from Research & Innovation Trasna Solutions Ltd. (Europe), *Reinforcement Learning for Irregular Slotted Aloha (IRSA) with short frames* by Iman Hmedoush from Nokia Bell Labs, France, *Evolution of network architectures and protocol stacks* by Constantine Dovrotis from Georgia Institute of Technology, USA, and *Cloud Solution Architect (Artificial Intelligence & Machine Learning)* by Franck Gaillard from Microsoft, France.

We would like to thank all who contributed to the success of this conference, in particular the members of the Program Committee and the reviewers for carefully reviewing the contributions and selecting a high-quality program. Our special thanks go to the members of the Organizing Committee for their great help.

We hope that all participants enjoyed this successful conference.

MLN 2022 was jointly organized by the EVA Project of INRIA Paris, the Laboratoire d'Informatique Gaspard-Monge (LIGM) and ESIEE Paris of Université Gustave Eiffel.

December 2022   Paul Mühlethaler
                                                                                      Éric Renault

Organization

General Chairs

Paul Mühlethaler   Inria, France
Éric Renault   ESIEE Paris, France

Steering Committee

Selma Boumerdassi   CNAM, France
Éric Renault   ESIEE Paris, France

Publicity Chair

Christophe Maudoux   CNAM, France

Organization Committee

Lamia Essalhi   ADDA, France

Technical Program Committee

Alberto Ceselli   Università degli Studi di Milano, Italy
Alberto Conte   Nokia Bell Labs, France
Antonio Cianfrani   University of Rome Sapienza, Italy
Anwer Al-Dulaimi   EXFO, Canada
Aravinthan Gopalasingham   Nokia Bell Labs, France
Cherkaoui Leghris   Hassan II University, Morocco
Claudio A. Ardagna   Università degli Studi di Milano, Italy
Éric Renault   ESIEE Paris, France
Eun-Sung Jung   Hongik University, South Korea
Fred Aklamanu   Nokia Bell Labs, France
Gianfranco Nencioni   University of Stavanger, Norway
Gloria Elena Jaramillo   DFKI, Germany
Hacène Fouchal   Université de Reims Champagne-Ardenne, France

Hella Kaffel Ben Ayed	Université de Tunis El Manar, Tunisia
Jean-Charles Grégoire	INRS, Canada
Jean-François Bercher	ESIEE Paris, France
Khaled Bousseta	Univ. Paris 13, France
Luiz Bittencourt	University of Campinas, Brazil
Mai Trang Nguyen	Univ. Paris 13, France
Mariam Kiran	Lawrence Berkeley National Laboratory, USA
Martine Wahl	Univ. Gustave Eiffel, France
Maxim Bakaev	NSTU, Russia
Miki Yamamoto	Kansai University, Japan
Mohamed Belaoued	Université 20 Août 1955-Skikda, Algeria
Mohamed Lahby	Université Hassan II, Morocco
Mounir Tahar Abbes	University of Chlef, Algeria
Nadjib Achir	University Sorbonne Paris Nord, France
Naïla Bouchemal	ECE, France
Nardjes Bouchemal	University Center of Mila, Algeria
Nicola Bena	Università degli Studi di Milano, Italy
Nirbhay Chaubey	Ganpat University, India
Olaf Maennel	Tallinn University of Technology, Estonia
Patrick Sondi	IMT Nord Europe, France
Paul Mühlethaler	Inria, France
Paulo Pinto	Universidade Nova de Lisboa, Portugal
Ruben Milocco	Universidad Nacional des Comahue, Argentina
Sabine Randriamasy	Nokia Bell Labs, France
Sherali Zeadally	University of Kentucky, USA
Smain Femmam	Université de Haute Alsace, France
Stavros Shiaeles	University of Portsmouth, UK
Van Khang Nguyen	Hue University, Vietnam
Van Long Tran	Enlab Software company, Vietnam
Viet Hai Ha	Hue University, Vietnam
Vinod Kumar Verma	SLIET, India

Sponsoring Institutions

ESIEE, Paris, France
Inria, Paris, France
Université Gustave Eiffel, France

Contents

Comparison of AI-Based Algorithms for Low Energy Communication

Morgane Joly[1], Éric Renault[2(✉)], and Fabian Rivière[1]

[1] NXP Semiconductors, 2 Esplanade Antone Philips, 14000 Caen, France
[2] LIGM, Univ. Gustave Eiffel, CNRS, ESIEE Paris, 93162 Noisy-le-Grand, France
eric.renault@esiee.fr

Abstract. Spectrum environment is more and more crowded by the presences of wireless sensor network (WSN), but radios still present an irregular quality of service (QoS) with a strong signal disturbances in their vicinity. Here, the main disturbance factors at the analogue level restricting the performances of radios using the 802.15.1 protocol are presented. The variety of solutions existing today, dummy as well as artificial intelligent (AI) based, are compared. A discussion of their advantages, their limiting conditions and their evolution is proposed.

Keywords: Cognitive Radio · Machine Learning · RF
Countermesures · BLE

1 Introduction

Wireless connections linking electronic devices are growing rapidly (Fig. 1) [7]. Built to increase user mobility, their exchange protocol does not need physical support allowing many field to use them as reliable tools and demonstrate the scalability of WSN devices. Thus, IoT is knowing a real democratisation in recent years; they are found in: medical field [19], automotive field [23], agriculture field [8], indoor geolocation [15], and, maybe the most known of large public, the personal devices that make technology increasingly modular with the large possibility of interconnections between them. The sensor sector is therefore creating a strong demand for wireless, short-range and low-speed technologies. Not to mention the MATTER standard developed by 28 companies which aim to rule all kind of IoT device [1] and which will certainly catalyse the boom of IoT devices.

IEEE 802.15 working group created standards to meet this need, which brings together all the standards relating to "Wireless Personal Area Networks" (WPAN). The standards concerned by this strong demand are Bluetooth Low Energy (BLE) that emit on the Industrial, Scientific and Medical (ISM) band and Zigbee which emit on ISM band and 868 MHz/915 MHz band. Among others, the most cited as interference is the wireless local area networks (WLAN) Fig. 2, which is the main user with powerful emission, and less common, waves escaping from microwave ovens [14]. The ISM band is therefore subject to local overloads due to the concentration of devices and exchanges. To ensure a minimum of

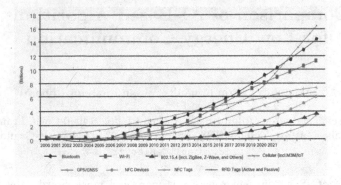

Fig. 1. IoT trend according Cisco white paper [7]

quality the IEEE 802.15.2 standard proposes a number of countermeasures and best practices:

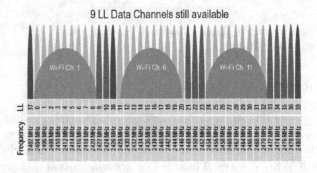

Fig. 2. Overlap of bluetooth low energy (BLE) by WLAN in ISM band

Based on fixed threshold tactics, these countermeasures lack flexibility (cf. 3) and are now struggling to maintain a good quality of service in overloaded frequency spaces where coexistence can only be achieved at the expense of the throughput of one of the networks: in the case where the WSN is the victim, for example during simultaneous transmission with a WLAN, the packet loss rate (PLR) can rise from 0% to 70% [4]. Thus, Elshabrawy [9] shows that interference caused by WLAN degrades large sensor networks. Similarly, [6] present the drastic drop in throughput in an overloaded WSN (up to 200 sensors) without WLAN (Fig. 3) and [22] show the augmentation of latency during device discovery procedure according to number of interfering device: up to a multiplication of more than 2.2 of the energy consumed (Fig. 4).

In spite of that, these interpretations can be confirmed with the study [16] which analyses the performance of a small Body Area Network (BAN). We

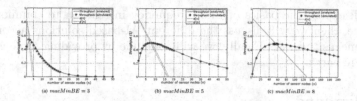

(a) $macMinBE = 3$ (b) $macMinBE = 5$ (c) $macMinBE = 8$

Fig. 3. Decrease of the throughput depending on the number of sensors in a WSN, with three different minimum back off periods (macMinBE) between two packets [6] in a context of saturated data traffic

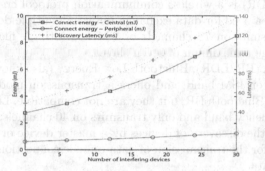

Fig. 4. Measurements for device discovery procedure [22]

observe, on the one hand, an increase in energy consumption between the optimal environment and the harsher environment (4 nodes under WLAN + BLE interference): consequence of multiple retransmissions; and, on the other hand, an increase in transmission latency and therefore a reduction in QoS (quality of service). So this shows that in terms of energy consumption BLE is resilient to interference from other technologies on small piconets[1].

It is therefore necessary to review the current coexistence strategies [2] to cleverly sobering the devices emissions in order to improve the resilience of IoT network. In other words, to make the radio intelligent (also called cognitive radio).

Since AI algorithms have become affordable in terms of computation and energy performance, a lot of research has been done in the field of cognitive radio to apply these algorithms to radios. Cognitive Radio explore the use data science and artificial intelligence (AI) with radio frequency data to provide flexible and adaptive solutions to multidimensional problems in a field where solutions would be long and costly to develop due to the phenomenal number of possible test cases

Since the research in Cognitive Radio (CR) are usually focused on Wifi or cellular communication major advances are rarely tested on BLE, that still a new technology. We will question the state of the art studies of CR and how they can be adapted to the inherent problematic of BLE.

---

[1] Networks composed by devices exchanging through BLE.

We will expose the basics of radio frequency in relation to common signal interference, then we will discuss the different AI-based solutions according to the radio means they use. Finally, we will conclude by providing some guidance for future research.

2   Background

2.1   Standard 802.15.1: Bluetooth Et Bluetooth Low Energy

Bluetooth (BR/EDR) is a wireless communication protocol created in 1994 by Ericsson. It meets a need for data exchange between multiple devices over short distances. It transmits on 79 channels of 2 MHz width. A master device can receive and manage a maximum of seven slaves.

Like Bluetooth (BR/EDR), Bluetooth Low Energy (also known as Bluetooth Smart) transmits on ISM band, and offers a transmission modulation method that is a cousin of Bluetooth BR, but they are not compatible. Like Bluetooth, it uses the ISM frequency band and only transmits on 40 channels: 3 "advertising" channels used for the discovery of devices by a master device or for the creation of connections or for the transmission of data outside connections and 37 for the transmission of data.

It meets more restrictive energy needs than BT thus everything is optimised for lowest power consumption:

– Data rate is about some kbits/s
– Small chunks of data transferred (exposing state)
– Short packets reduce RX time and peak current
– Short radio communication range
– Single protocol : Broadcast, Connection, Event Data Models, Reads, Writes
– Less RF channels to improve discovery and connection time

It also supports a wide range of network topologies (point-to-point, broadcast, mesh) to enable the interconnection of networks required for the development of the Internet of Things. The number of slaves managed by a master is limited only by the application.

2.2   Limiting Factors in a Receiver (analogue Part)

Here a reminder of major physical sources which impact the quality of the received signals.

Thermal Noise. Thermal noise increases in presence of interferer due to AGC protecting receiver from saturation. Interferer degrades sensitivity so receiver range. It is due to the movement of electrons in the material. This noise is considered random and constant so it determines the sensitivity floor of a receiver: in ideal conditions (no interferer) the signal must exceed the power of the thermal noise to be perceptible by the receiver.

Phase Noise Due to Un-Ideal Local Oscillator (LO) Signal. The local oscillator bring the desired signal around 0 Hz frequency (suppression of the carrier). Due to residual signals around the target frequency it create a skirt around the carrier frequency f0 in frequency domain and translate noise in addition of wanted signal The height of the skirt is controlled by the energy supplied to the oscillator: the more energy used, the lower the skirt and the less the interferer will corrupt the wanted signal, but the more energy the radio will consume (Fig. 5).

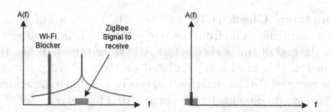

Fig. 5. Impact of local oscillator on interferers around the wanted signal

2nd **and** 3nd **order intermodulation products** Amplifier non-linearities create intermodulation products components proportional to interferes level during the amplification by the low noise amplifier (LNA) and which will overlap to the desired signal.

Gain Compression. The final source of significant sensitivity loss is signal compression. Receiver output signal gain drops and approaches zero for sufficiently high input level. While this degradation has little impact on frequency-modulated signals (the signal transmits information by frequency variation), it causes significant information loss in amplitude-modulated signals (the signal transmits information by amplitude variation).

2.3 Evaluate the Quality of the Link

Estimating the quality of the link is the first step in observing the network. Thus, their selection as an input parameter in an algorithm is very important.

Fast Link Estimators: Received Signal Strength Intencity (RSSI) and Link Quality Indicator (LQI). They are two complementary metrics of link quality in stimulated wireless links RSSI and LQI are two hardware-based complementary metrics of link quality between two radios. The RSSI produces an estimate of the average power in a channel throughout the reception of a packet, all signals included, while the LQI quantifies the signal-to-noise ratio, i.e. the capacity of the receiver to decode the signal despite ambient noise. They can therefore be used to quickly estimate the quality of a link, but are subject to various biases that make them a poor resource for determining whether a signal has been corrupted [5].

Signal to Noise Ration (SNR). The SNR gives the margin of the signal over the noise and the ability of the demodulator to recognize the bits.

Cyclic Redundancy Check (CRC). The CRC is a word of some bit calculated by the transmitting radio from the payload part of the transmitted packet. It is added to the end of the packet before it is transmitted. When the receiver decodes the packet, this word is recalculated on the received payload and compared with the received CRC. From this, the receiver is able to determine whether the received packet corresponds to the transmitted packet.

3   Usual Countermeasures

Carrier Sense Multiple Access (CSMA). Carrier Sense Multiple Access is an unsupervised coexistence technique used by 802.11 and 802.15.4. All nodes must listen on the channel. The first packet in the transmission queue is sent when the detection result mentions an idle channel. Each node has the same chance of access.

Time Division Multiple Access (TDMA). This supervised coexistence mean divides the channels into several real time slots. Each time slot is allocated to a different transmitter. The allocation is done by a base station thus needs additional devices to function well Spectrum usage and requirements are predictable.

Frequency Hopping (FH). Frequency hopping is a coexistence technique that works very well with TDMA. The sender and the transmitter agree on a calculation to change channels at each packet transmission.

Automatic Gain Control (AGC). To reduce the impact of saturation the Automatic Gain Control adjusts the receiver gain according to the received signal strength to avoid saturation within the receiver. It converges when the packet preamble is received. It allows the radio to adapt only to the power of the beginning of the packet. However if the interference arrives after the freeze of the AGC index then the signal will be saturated and the packet will be corrupted. The packet re-emission procedure will be triggered, causing an over consumption of energy compared to a packet well received at the first try.

4 AI-Based Countermesures for Cognitive Radio

In the Sect. 3 we saw that usual counter measure does not give awareness of the environment to the radio.

4.1 Spectrum Prediction

Spectral prediction allow to anticipate the evolution of spectrum congestion.

In the study [21], Sudharsan et al. propose an adaptive layer that predict the RSSI evolution thanks to a Support Vector Regression (SVR) to guide the choice of the best protocol to use in the future environment condition. Thanks to its predictions, the Adaptive Strategy Block (ASB) is able to maintain an optimal quality of service. This method add some flexibility in the protocol with a solution independent of the platform. Nevertheless, it implies integrating in the hardware all the possible protocol to allow a device to have the best choice panel. The device must therefore be able to support the energy consumption of the most energy-intensive protocols (Wi-Fi) in addition to being expensive.

Although the RSSI is not a sufficient parameter to characterize the strength of a link, this study indicates that it is possible to predict the strength of a link precisely, over a period of about one minute. But the conducted experimentation are based on continuous dataset without taking into account the frequency hopping. This imply that the RSSI registered will not correspond to the same channel so the same environment. Therefore, in real condition, the algorithm will make its prediction on partial data and the real accuracy will be impacted. To ensure that the predictions of this algorithm will be usable on all channels, the prediction must remain very accurate over a sufficiently long time window to be usable when the channel is to be active again. And this obviously implies an additional memory allocation to retain all the predictions made for each channel.

4.2 Waveform Classification of Interfering Technologies

The recognition of the technologies in the vicinity of the radio is an important asset for a radio system. Indeed, their countermeasures behaviour can be adapted to the specific pattern of the interference produced by a protocol.

Thus, Grimaldi et al. [11] propose the real-time identification of three standards (IEEE 802.11.b/g/n, IEEE 802.15.1 and IEEE 802.11.2) by four different machine learning algorithms (two types of classification tree, namely CT1 and CT2, a Random Forest of Classification Trees (RFCT) and a Multiclass-SVM (MSVM)). Because filtering methods degrade the spectral signatures of interference, the team chose to use the RSSI from the raw signals. It extracts, on several frequencies surrounding the channel, the physical characteristics of the interference (length of the energy peak, average power, maximum variation of the envelope, measurement of the variation of the maximum power between two samples) allowing a greater precision on the acquisition of the samples necessary for their algorithm (Fig. 6).

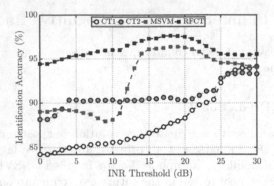

Fig. 6. Accuracy comparison of the four algorithms proposed by Grimaldi and al. [11]

As for Zhang et al. [24] to perform their standard identification, they transform the signal received from a channel into a 2D image relating time and frequencies in order to exploit the performance of Convolutional Neural Networks (CNN) for their classification. While [17] choose to use the temporal characteristics and phase of the signals with a Long-Short Time Memory (LSTM) algorithm. Whatever the algorithms used, it is notable that it is the choice and pre-processing of the data that allows good accuracy to be obtained in the tests.

Although promising, the main limitation of this technique lies in the multiplicity of waveforms. The algorithm has to be retrained for each new technology and the teams' tests are only done on single interefer type at a time, whereas, in the same location, many technologies can coexist, which must qualify the performance of these algorithms.

4.3 Temporal Location of Intra-package Damage

Determining the causes of packet loss ultimately allows the most effective countermeasures to be triggered based on a known problem.

The study [10] shows that it is possible to differentiate the events that may have caused the loss of a packet (weak signal, partial temporal overlap of interference, total overlap of interference). Using probabilistic probing they distinguish wireless losses due to signal weakness from losses due to multiple access collisions. In a BLE radio system, this countermeasure would certainly require the support of other parameters to implement an efficient spectrum evolution prediction strategy leading to a delay of connection events or a frequency hopping to reduce the number of interfered bits in a packet.

The team [18] focuses on the duty cycles adopted by Wi-Fi according to the activities of users. By analysing the characteristics of received packets, they distinguish interference due to web browsing, video viewing or Skype calls. With this analysis, it is possible to design a system capable of identifying the activities of a competing network and to propose an efficient connection offset.

The differentiation of the types of interference impact would allow, for example, to propose an efficient connection offset to allow the master-slave pair to synchronise their exchanges between WLAN bursts or, as realised by Hanna et al. [12], to facilitate the construction of the bit-by-bit signal to reduce latency (by reducing the number of packet retransmissions) and power consumption.

5   Futur Work

In these both study [3, 20], research have been made on improving channel allocation for licensed band, the challenge lies in the ability of a secondary user to use the spectrum without interfering with the transmission of a primary user. To our knowledge, no such work has been done on the ISM band with BLE or Zigbee, which would be very interesting in terms of predicting channel spectrum congestion for each channel.

Hithnawi's thesis [13] proposes an improvement to the error detection algorithm using the Hamming distance and RSSI with empirical thresholds. It does not use an AI algorithm but the improvements in error detection are very promising. The Hamming distance quantify of noise immunity, so it reflects the confidence in symbol decoding for each decoded symbol, it will be quite eligible as a parameter in an ML algorithm.

As mention in 3, the AGC only protect the BLE radio from saturation using the signal power at the beginning of the packet reception without taking into account the late arriving of a interferer during the reception. This late saturation could be used to make a prediction on the future channel congestion or the make a pattern recognition to adapt the connection event of the BLE radio.

6   Conclusion

Actual cognitive radio research are not focused on the 802.15.1 protocol but can be adapted in full knowledge of its specificity. The three major to improve radio QoS are spectrum prediction, classification of surrounding technologies and identification and correction of intra-packet errors.

We have also seen that the amelioration and utilisation of actual countermeasures proposed by the standards can be used to develop new ML-based countermeasures.

References

1. Matter Project. https://github.com/project-chip/connectedhomeip/readme
2. IEEE recommended practice for information technology- local and metropolitan area networks- specific requirements- part 15.2: Coexistence of wireless personal area networks with other wireless devices operating in unlicensed frequency bands. IEEE Std 802.15.2-2003, pp. 1–150 (2003). https://doi.org/10.1109/IEEESTD.2003.94386

3. Alhammadi, A., Roslee, M., Alias, M.Y.: Analysis of spectrum handoff schemes in cognitive radio network using particle swarm optimization. In: 2016 IEEE 3rd International Symposium on Telecommunication Technologies (ISTT), pp. 103–107 (2016). https://doi.org/10.1109/ISTT.2016.7918093
4. Angrisani, L., Bertocco, M., Fortin, D., Sona, A.: Experimental study of coexistence issues between IEEE 802.11b and IEEE 802.15.4 wireless networks. IEEE Trans. Instrument. Measur. **57**(8), 1514–1523 (2008). https://doi.org/10.1109/TIM.2008.925346
5. Barać, F., Gidlund, M., Zhang, T.: Ubiquitous, yet deceptive: Hardware-based channel metrics on interfered WSN links. IEEE Trans. Vehicul. Technol. **64**(5), 1766–1778 (2015). https://doi.org/10.1109/TVT.2014.2334494
6. Cao, X., Chen, J., Sun, Y., Shen, X.: Maximum throughput of IEEE 802.15.4 enabled wireless sensor networks. In: 2010 IEEE Global Telecommunications Conference GLOBECOM 2010, pp. 1–5 (2010). https://doi.org/10.1109/GLOCOM.2010.5683611
7. Cisco: IEEE 802.11ax: The sixth generation of wi-fi white paper (April 2020). Accessed 13 May 2022
8. de la Concepcion, A.R., Stefanelli, R., Trinchero, D.: Adaptive wireless sensor networks for high-definition monitoring in sustainable agriculture. In: 2014 IEEE Topical Conference on Wireless Sensors and Sensor Networks (WiSNet), pp. 67–69 (2014). https://doi.org/10.1109/WiSNet.2014.6825511
9. Elshabrawy, T.: Throughput analysis of IEEE 802.15.4 enabled wireless sensor networks under wlan interference. In: 2014 IEEE Fourth International Conference on Consumer Electronics Berlin (ICCE-Berlin), pp. 467–469 (2014). https://doi.org/10.1109/ICCE-Berlin.2014.7034322
10. Eu, Z.A., Lee, P., Tan, H.P.: Classification of packet transmission outcomes in wireless sensor networks. In: 2011 IEEE International Conference on Communications (ICC), pp. 1–5 (2011). https://doi.org/10.1109/icc.2011.5962637
11. Grimaldi, S., Mahmood, A., Gidlund, M.: Real-time interference identification via supervised learning: Embedding coexistence awareness in IoT devices. IEEE Access **7**, 835–850 (2019). https://doi.org/10.1109/ACCESS.2018.2885893CNN
12. Hanna, S., Dick, C., Cabric, D.: Combining deep learning and linear processing for modulation classification and symbol decoding. In: GLOBECOM 2020–2020 IEEE Global Communications Conference, pp. 1–7 (2020). https://doi.org/10.1109/GLOBECOM42002.2020.9348060
13. Hithnawi, A.: Low-power Wireless Systems Coexistence. Diss. ETH No. 23907 (2016)
14. Baylon, J., Ethan Elenberg, S.M.: ISCISM: Interference sensing and coexistence in the ISM bandj. High Frequency Design (2012)
15. Kolakowski, M.: Automatic radio map creation in a fingerprinting-based ble/uwb localisation system. IET Microwaves Anten. Propag. **14**(14), 1758–1765 (2020). https://doi.org/10.1049/iet-map.2019.0953
16. La, Q.D., Nguyen-Nam, D.V., Ngo, M., Hoang, H., Quek, T.Q.: Dense deployment of ble-based body area networks: A coexistence study. IEEE Trans. Green Commun. Netw. 1 (2018). https://doi.org/10.1109/TGCN.2018.2859350
17. Rajendran, S., Meert, W., Giustiniano, D., Lenders, V., Pollin, S.: Deep learning models for wireless signal classification with distributed low-cost spectrum sensors. IEEE Trans. Cognit. Commun. Netw. **4**(3), 433–445 (2018). https://doi.org/10.1109/TCCN.2018.2835460
18. Palit, R., Kshirasagar Naik, A.S.: Anatomy of wifi access traffic of smartphones and implications for energy saving techniques

19. Rivero-Angeles, M., et al.: Mobile clustering scheme for pedestrian contact tracing: The covid-19 case study. Entropy **23** (2021). https://doi.org/10.3390/e23030326
20. Sengottuvelan, S., Ansari, J., Mähönen, P., Venkatesh, T., Petrova, M.: Channel selection algorithm for cognitive radio networks with heavy-tailed idle times. IEEE Trans. Mob. Comput. **16**(5), 1258–1271 (2017). https://doi.org/10.1109/TMC.2016.2592917
21. Sudharsan, B., Breslin, J.G., Ali, M.I.: Adaptive strategy to improve the quality of communication for IOT edge devices. In: 2020 IEEE 6th World Forum on Internet of Things (WF-IoT), pp. 1–6 (2020). https://doi.org/10.1109/WF-IoT48130.2020.9221276
22. Treurniet, J.J., Sarkar, C., Prasad, R.V., De Boer, W.: Energy consumption and latency in ble devices under mutual interference: An experimental study. In: 2015 3rd International Conference on Future Internet of Things and Cloud, pp. 333–340 (2015). https://doi.org/10.1109/FiCloud.2015.108
23. Yao, W., Pang, Z., Zhuang, K., Shao, W.: Design and application of wireless temperature monitoring system for diesel locomotive in reconditioning field based on zigbee network. In: Jia, Y., Zhang, W., Fu, Y. (eds.) Proceedings of 2020 Chinese Intelligent Systems Conference, pp. 470–478. Springer, Singapore (2021)
24. Zhang, M., Diao, M., Guo, L.: Convolutional neural networks for automatic cognitive radio waveform recognition. IEEE Access **5**, 11074–11082 (2017). https://doi.org/10.1109/ACCESS.2017.2716191

Development of an Intent-Based Network Incorporating Machine Learning for Service Assurance of E-Commerce Online Stores

Remigio Hurtado[✉], Mario Torres, Bryan Pintado, and Arantxa Muñoz

Universidad Politécnica Salesiana, Calle Vieja 12-30 y Elia Liut, Cuenca, Ecuador
rhurtadoo@ups.edu.ec, {ltorresg3,bpintadoy}@est.ups.edu.ec

Abstract. Managing the large number of network devices is a major challenge for organizations. The management of these devices requires meticulous care since manual configurations of traditional networks often lead to errors. Manual configurations result in inefficient troubleshooting. Instead of worrying about the development of new services, the organizations' staff is exhausted by the arduous task of maintaining network devices. Specifically, one of the organizations' core services is their e-commerce service available through their website, through which they present and sell products. Companies that have had a great growth require optimizing the use of CPU, memory, bandwidth resources, among others, in order to provide an available and scalable service for the large number of customers. Due to this, this paper proposes: 1) an Intent Based Network (IBN) based on Cisco's IBN architecture applied to online stores in e-commerce, 2) a scalable and efficient method that uses Neural Networks (NN) for resource prediction in order to adapt to the large number of customers and the growing demand in this era of Big Data. The IBN architecture provides automation, flexibility and unlike Software Defined Networks (SDN) includes the integration of artificial intelligence and data mining techniques in order to analyze future trends and problems for network service assurance. The neural network predicts the appropriate bandwidth value based on the trend of the number of users that the online store server will consume and other resources such as CPU, memory and storage usage. This value is automatically configured in the network, so that the service always has a high availability. We have created a dataset by collecting data from the network we have developed. We have defined a base neural network model and subsequently optimized this model by minimizing the Mean Absolute Error (MAE). We have shown that the optimized neural network outperforms other machine learning methods such as Random Forest (RF), Support Vector Machine (SVM) and K Nearest Neighbor (KNN). This research opens the door to the development of IBNs incorporating artificial intelligence for e-commerce and other services.

Keywords: intelligent networks · intent based networks · artificial intelligence · machine learning · neural networks · online stores · website · e-commerce

E. Renault and P. Mühlethaler (Eds.): MLN 2022, LNCS 13767, pp. 12–23, 2023.
https://doi.org/10.1007/978-3-031-36183-8_2

1   Introduction

The importance and effect of day-to-day networking and communication systems is immense, such that every minute of daily life is influenced in some way by smartphones and smart devices [1]. This essential transformation of human society relies on increasingly advanced communication networks in order to achieve near-instantaneous transport of massive amounts of data from billions of connected devices [2]. In [3] it is mentioned that e-commerce is a reality present in people's daily lives, and with this the problems that come with having many devices trying to connect and perform electronic transactions. Managing the large number of network devices is a major challenge for organizations. The management of these devices requires meticulous care since manual configurations of traditional networks often lead to errors. Manual configurations result in inefficient troubleshooting. Instead of worrying about the development of new services, the organizations' staff is exhausted by the arduous task of maintaining network devices. Specifically, one of the organizations' core services is their e-commerce service available through their website, through which they present and sell products. Companies that have had a great growth require optimizing the use of CPU, memory, bandwidth resources, among others, in order to provide an available and scalable service for the large number of customers.

Since 2015, Intent-Based Network (IBN) was proposed as a new network management framework [4] and, it is a promising solution as it can manage and control network configurations, modifications and operation processes in an automated way. It is an intelligent network system that can convert, deploy and configure network resources according to the operator's intentions automatically [5]. In the IBN approach, a network should have several characteristics, such as self-organization, self-reliance, self-healing, and self-configuration [6]. One of the most challenging research directions in communication networks is to build autonomous networks, i.e., networks that can operate with little or no human involvement. However, today's communication networks can benefit greatly from autonomous operation and adaptation, not only because of the implied cost savings, but also because autonomy will enable functionalities that are unfeasible today [7]. Intent-based networks offer to simplify network management and automated orchestration of high-level policies in network architectures such as Software Defined Networking (SDN) [8]. An example is that an expert can declaratively specify a network intent with a policy of "approve hosts X and Y to communicate with a bandwidth capacity Z" without having to understand the details of the mechanism over the switch, forwarding configurations, or topology [9]. That is, IBNs can extend the programmable functionality found in SDNs by allowing practitioners to specify what policies they want their network to implement rather than how the underlying mechanisms of their network will implement those policies. Therefore, this approach enables simplification and abstraction of complex network management [10]. In [12] it is mentioned that Gartner predicted that IBNs would be the big thing in the networking field. Compared to traditional network management solutions, IBN has shown great potential in numerous aspects, reducing network configuration implementation time and improves network scalability [11].

IBNs provide four capabilities: 1) Translation and validation, 2) Automated deployment, 3) Network state awareness, and 4) Dynamic assurance and optimization. In a general way, the process of an IBN is as follows: IBNs collect intentions (requirements, needs and network data), these intentions are translated and validated by means of the organization's policies. The definitions of intent in each working group are slightly different and specific to their respective core technology. However, intent is understood as a high-level statement of the objectives, service level or business objectives that the network should be able to achieve or the desired behavior that the network is expected to accomplish. The intent does not entail concrete actions or configuration steps that the network should take to achieve the described objectives. Humans are prone to make mistakes and the user can enter a wrong intention, in this phase, a validation of the entered policy will be performed and in turn it can be executed. Subsequently, these policies are automatically executed by a controller in the network infrastructure. The network is constantly monitored, data is collected to analyze trends and anticipate problems. Trends and problems are the new intentions of the network that must be dealt with by the intelligent network automatically. This whole process is cyclical, so that the network is autonomous and intelligent. An important feature to note about IBN is adaptability to changes in scale, as mentioned in [13], this is a fundamental property of a well-designed IBN system, requiring the ability to consume and process analysis methods that are context/intent aware at near real-time speeds. Cisco defines that intent can be applied to application service levels, security policies, compliance, operational processes, and other business needs. Intent-based networks capture and translate business intent so that it can be applied throughout the network [12].

In an IBN architecture, one or more centralized controllers are responsible for managing different segments of the target area, and each segment runs on a virtualized network infrastructure [14]. It is mentioned that IBN aims to meet massive demands for intelligent services and overcome variations over time. IBNs can continuously learn and adapt to the time-varying network environment based on massively collected real-time network data. This makes use of large amounts of data that are collected in real time so that the network can learn and adapt according to user demand. For descriptive and predictive learning and analysis, it is necessary to incorporate artificial intelligence techniques, specifically machine learning techniques. Machine Learning (ML) is the branch of artificial intelligence that aims to develop techniques that allow computers to learn. More specifically, it is about creating algorithms capable of generalizing behaviors and recognizing patterns from information provided in the form of samples [16]. ML is the technology of developing computer algorithms that are capable of emulating human intelligence [17].

Due to the problems of traditional networks and the many benefits that IBNs provide, this research proposes: 1) an intent based network applied to online stores in e-commerce, 2) a scalable and efficient method that uses neural networks for resource prediction in order to adapt to the large number of customers and the growing demand in this era of Big Data. The IBN architecture provides automation, flexibility and unlike SDN includes the integration of artificial intelligence and data mining techniques in order to analyze future trends and problems for network service assurance. The neural network predicts the appropriate bandwidth value based on the trend of the number of users that the online store server will consume. This value is automatically configured in the network, so that the service always has a high availability.

The **most important contributions** of this work are the following:

- An intent based network applied to online stores in e-commerce
- A scalable and efficient method that uses neural networks for resource prediction in order to adapt to the large number of customers and the growing demand in this era of Big Data.
- A process of learning and optimization of the neural network in order to obtain the best prediction results.
- A set of experiments to demonstrate that the optimized neural network outperforms other regression problem-oriented machine learning methods, such as random forest, support vector machine and k nearest neighbor.

The rest of this paper is structured as follows: Sect. 2 mentions the most relevant related work, Sect. 3 presents the methodology used, Sect. 4 presents the design of experiments. Subsequently, in Sect. 5, the results and discussion are presented. Finally, Sect. 6 presents the conclusions and future work from this research.

2 Related Work

This section presents some relevant papers related to this research.

In the research of [19], an efficient solution that automates the configuration process and performs network segment management and orchestration is proposed. This solution contains an IBN network platform that effectively organizes and manages the lifecycle of multi-domain segment resources. In the work of [20], it is shown that by implementing an automation methodology, virtual local area networks (VLANs) can be configured by passing parameters directly through programmed code. The fundamentals and the most relevant machine learning algorithms are explained in [16]. Neural networks and other relevant techniques such as support vector machines and ensemble algorithms are the techniques that generate the best results in supervised problems. However, with the era of Big Data, neural networks through deep learning have a bright future. In the work of [15], the strength of certain methods for analyzing sequences of variable size is demonstrated by their unfolding in time as a function of the size of the input. Specifically, bidirectional recurrent neural networks are constructed, as a specification of recurrent networks, showing their potential in non-causal systems, where inputs may depend on inputs from past and future times. In addition, a platform is developed to implement dynamic recurrent networks, with a backpropagation trough time learning algorithm to allow the development of networks for any problem where the inputs are sequences analyzed in time and the output are other sequences or simply descriptors of functions or properties of these.

3   Proposed IBN Architecture and Machine Learning Method

This section first presents the IBN architecture designed for a online store environment, and then the proposed method to perform the predictive analysis.

3.1   IBN Architecture

Figure 1 shows the proposed architecture of the intent-based network, which consists of four phases: 1) Translation (of the Intent), 2) Activation (according to access policies for network automation), 3) Physical and virtual infrastructure, 4) Assurance (data collection and monitoring, prediction with the machine learning model. These phases are repeated iteratively in the same sequence. Each phase is described below:

- **Phase 1 - Translation:** in this phase the intent is interpreted, network operators are authorized to express the intent in an explanatory and manageable way, expressing what the network's attitude is expected to be in order to protect the business objectives and achieve the desired performance.
- **Phase 2 - Activation:** Once the intent is captured, it is then translated into policies so that they can be automatically employed in the network. In this phase, policies are instantiated in the virtual and physical network infrastructure through network automation.
- **Phase 3 - Physical and virtual infrastructure:** in this phase, configurations are applied to network devices, including routers, switches and virtual machines.
- **Phase 4 - Assurance:** to verify that the network behaves as intended, machine learning is performed to constantly monitor the network, analyze patterns and trends, anticipate problems, and consequently help achieve business objectives.

To monitor the network, information is retrieved about the running processes and the utilization of system resources, in this case, CPU, memory, storage, number of website visits, and bandwidth. The tools and libraries used are: Python, PSUTIL and SPEEDTEST. Figure 2 shows the network topology, incorporating

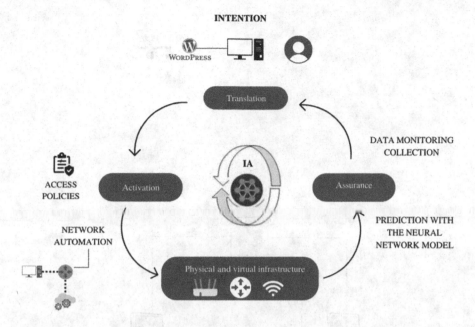

Fig. 1. Proposed IBN architecture for an e-commerce online store environment

the web server, network devices and end devices. Requests to the server are simulated using the Apache JMeter tool. The data is collected and serves as input to the machine learning model. The model predicts the bandwidth, and this data according to its value (if the trend is a problem for the network) becomes a new intent for the IBN network. Request simulation, learning and prediction are performed by the Performance Test Host. The IBN network translates this intent and automatically configures the network. The new network state is monitored to ensure proper web server behavior. The network is constantly monitored for self-configuration.

3.2 Machine Learning Method Applied to the IBN Network

Figure 3 shows the process for training and optimizing the learning model. We have used a neural network, since it is a scalable method for a Big Data environment and the predictions are very fast once the model has been trained. The process is as follows: 1) Data collection from the web server (current bandwidth, amount of visits, cpu usage, total memory and memory used, total storage and storage used), 2) Base training of the neural network model with the Training set, 3) Optimization of the model by fine-tunning of compilation hyperparameters, such as: activation functions, number of epochs, optimizers, and batch

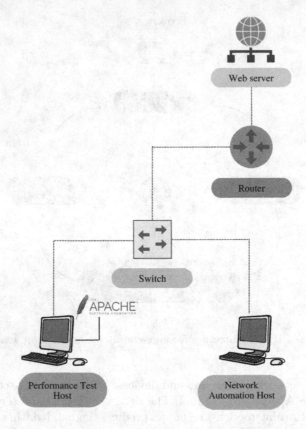

Fig. 2. Network topology

size, 4) Layer density optimization, 5) Model evaluation with the Tet set, 6) Bandwidth prediction according to the data collected at an instant of time, 7) Automatic network configuration and assurance. To determine the new bandwidth in order to ensure the virtual store service (thus giving a better loading time to the server), the equation (Eq. 1) is applied. For training, optimization and evaluation we use the MAE quality metric (see equation Eq. 2), which indicates the model error. We have incorporated data preprocessing in order to clean the noise (missing values) and normalize the data.

$$\text{extended burst} = \text{normal burst} * 2 \tag{1}$$

where normal burst (current bandwidth) is the data transfer of traffic channels and extended burst is the double the data transfer of the traffic channels.

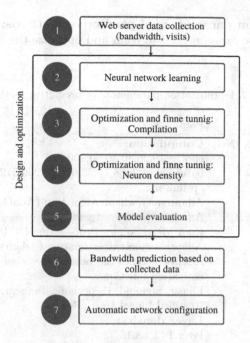

Fig. 3. Learning process and optimization of the machine learning method

4   Design of Experiments of Machine Learning

This section presents the characteristics of the dataset, the parameters for optimization of the machine learning methods in the learning phase, and the quality measure.

1) **Characteristics of dataset:** Table 1 presents the general description of datasets and their most relevant characteristics.

Table 1. Description of dataset

Dataset	Number of samples	Number of variables	Resources
Dataset collected	10000	7	current bandwidth (in megabytes)
			amount of visits
			cpu usage (in %)
			total memory (in gigabytes)
			memory used (in gigabytes)
			total storage (in gigabytes)
			storage used (in gigabytes)

2) Optimization parameters of predictive methods: Table 2 presents the parameters to be experimented to train and optimize the methods.

Table 2. Optimization parameters of predictive methods

Method	Parameters
Neural Network (NN)	**Compilation:** epochs:10,20,30,40,50,100 batch size:1,2,4,8,16,32,64,128,256,512 optimizers: Adaptive Gradient Algorithm (AdaGrad), Adadelta (AdaGrad variant), Root Mean Square Propagation (RMSprop), Adaptive moment estimation (Adam). Adam combines the best of the other optimizers. activation functions: Linear, Sigmoid, Hyperbolic Tangent, Rectified Lineal Unit (ReLU) **Layer density:** layer 1: 2,4,8,16,32 layer 2: 2,4,8,16,32 output layer: 1
Random Forest (RF)	number of estimators: 5,10,15,20,30,40,50 maximum depth: 5,10,15,20,40
SVM	penalty (C): 0.001, 0.01, 0.1 kernel: linear, polynomial, sigmoid, radial basis function (rbf) epsilon: 0.0,0.1,0.2,0.3,0.5,1
KNN	number of neighbors: 5,10,20,30,40,50,100,200 metrics: euclidea, manhattan

3) Quality measures: For the evaluation of the methods, the Cross-Validation K-Folds technique (with K=20) has been used in order to obtain an adequate generalization of the results. The quality measures used is: MAE. The average of K experiments with the best parameters of each method is presented in the results section.

$$\text{MAE} = \frac{\sum_{i \in Q} |r_i - p_i|}{\#Q} \tag{2}$$

where r_i is the actual test bandwidth, p_i is the predicted bandwidth and $\#Q$ is the number of samples in the Test set.

5 Results

This section presents the results of the machine learning methods and the results of applying machine learning to the IBN network.

The final parameters of the learning methods are presented in Table 3. Table 4 shows the comparison of the results of the machine learning methods. It can be seen that the optimized neural network outperforms the other machine learning methods. Figure 4 shows a heat map of the Grid Search optimization process for the neural network layer density optimization. It can be noticed that the best combination is L1 with 32 neurons and L2 between 4 and 32 neurons. With more than two intermediate layers, the results do not improve and the learning time increases. Therefore, an important decision is to determine the architecture of the neural network with two layers.

Table 3. Final parameter values

Method	Parameters
NN	**Compilation:** epochs: 50 batch size: 128 optimizer: Adam **Layer density:** layer 1: 32 (activation function - ReLU) layer 2: from 4 to 32 (activation function - ReLU) output layer: 1 (activation function - linear)
RF	number of estimators: 30 depth: 20
SVM	C: 0.01 kernel: linear epsilon: 0.2
KNN	number of neighbors: 20 metric: euclidea

To demonstrate the efficiency of our network, we have disabled the network automation and simulated 800 requests to the web server with the Apache JMeter tool. The average response time (performance without machine learning) of the server is about 17 s, which is an inadequate time for an online store. Now, then we have activated the network automation, the network identifies the problem as an intention, therefore, the neural network model predicts the bandwidth, and the IBN network automatically configures the new bandwidth. The new network state is monitored to ensure proper web server behavior. The average response time (performance with machine learning) is now 150 milliseconds, which is adequate. Google currently considers that the response time of a server should be below 200 milliseconds (0.2 s).

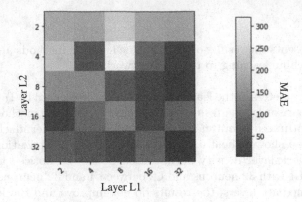

Fig. 4. Grid Search map of the optimization with two layers L1 and L2 in the proposed method by applying neural networks in the learning phase

Table 4. Comparison of results: machine learning methods

Machine learning method	MAE
NN	**5.5103**
RF	10.1443
SVM	15.1295
KNN	28.1671

6   Conclusions

Managing the large number of network devices is a major challenge for organizations. Manual configurations result in inefficient troubleshooting. Companies that have had a great growth require optimizing the use of CPU, memory, bandwidth resources, among others, in order to provide an available and scalable service for the large number of customers. With this research we have provided an Intent Based Network (IBN) based on Cisco's IBN architecture applied to online stores in e-commerce, a scalable and efficient method that uses neural networks for resource prediction in order to adapt to the large number of customers and the growing demand in this era of Big Data. The IBN architecture provides automation, flexibility and includes the integration of machine learning in order to analyze future trends and problems for network service assurance. Specifically, in this research, we have developed a neural network that predicts the appropriate bandwidth value based on the trend of the number of users that the online store server will consume and other resources. This value is automatically configured in the network, so that the service always has a high availability. This research opens the door to the development of IBNs incorporating machine learning methods for e-commerce and other services. The large amount of data and users on the Internet requires efficient techniques that are able to meet the new demands of the networks.

References

1. John, H.M.: Reclaiming What we've Lost in a World of Constant Connection. Penguin, The end of absence (2014)
2. Gary, D.: 2020: Life with 50 billion connected devices. In: 2018 IEEE International Conference on Consumer Electronics (ICCE). IEEE (2018)
3. Meryem, S., et al.: 5G-enabled tactile internet. IEEE J. Sel. Areas Commun. **34**(3), 460–473 (2016)
4. Yoshiharu, T., Okabe, Y.: Reactive configuration updating for intent-based networking. In: 2017 International Conference on Information Networking (ICOIN). IEEE (2017)
5. Khizar, A., et al.: Slicing the core network and radio access network domains through intent-based networking for 5G networks. Electronics **9**(10), 1710 (2020)
6. Khizar, A., et al.: Network slice lifecycle management for 5G mobile networks: an intent-based networking approach. IEEE Access **9**, 80128–80146 (2021)
7. Paul, H., et al.: Evolutionary autonomous networks. J. ICT Stand. **9**(2), 201–228 (2021)
8. Ujcich, B.E., Bates, A., Sanders, W.H.: Provenance for intent-based networking. In: 2020 6th IEEE Conference on Network Softwarization (NetSoft). IEEE (2020)
9. Davide, S., et al.: ONOS intent monitor and reroute service: enabling plug&play routing logic. In: 2018 4th IEEE Conference on Network Softwarization and Workshops (NetSoft). IEEE (2018)
10. Yoonseon, H., et al.: An intent-based network virtualization platform for SDN. In: 2016 12th International Conference on Network and Service Management (CNSM). IEEE (2016)
11. Engin, Z., Turk, Y.: Recent advances in intent-based networking: a survey. In: 2020 IEEE 91st Vehicular Technology Conference (VTC2020-Spring). IEEE (2020)
12. Lei, P., et al.: A survey on intent-driven networks. IEEE Access **8**, 22862–22873 (2020)
13. Tim, S., et al.: Cisco Digital Network Architecture: Intent-Based Networking for the Enterprise. Cisco Press, Indianapolis (2018)
14. Wei, Y., Peng, M., Liu, Y.: Intent-based networks for 6G: insights and challenges. Digital Commun. Netw. **6**(3), 270–280 (2020)
15. Bonet, C.I., et al.: Redes neuronales recurrentes para el análisis de secuencias. Revista Cubana de Ciencias Informáticas **1**(4), 48–57 (2007)
16. Bishop, C.M., Nasrabadi, N.M.: Pattern Recognition and Machine Learning. vol. 4. No. 4. New York: Springer (2006)
17. El Naqa, I., Murphy, M.J.: What is machine learning? In: El Naqa, I., Li, R., Murphy, M.J. (eds.) Machine Learning in Radiation Oncology, pp. 3–11. Springer, Cham (2015). https://doi.org/10.1007/978-3-319-18305-3_1
18. Ethem, A.: Introduction to Machine Learning. MIT Press, Cambridge (2020)
19. Khizar, A., et al.: Network slice lifecycle management for 5G mobile networks: an intent-based networking approach. IEEE Access **9**, 80128–80146 (2021)
20. Bello, L.L., Steiner, W.: A perspective on IEEE time-sensitive networking for industrial communication and automation systems. Proc. IEEE **107**(6), 1094–1120 (2019)

Cyber-attack Proactive Defense Using Multivariate Time Series and Machine Learning with Fuzzy Inference-based Decision System

Rahmoune Bitit[1]([⊠]), Abdelouahid Derhab[2], Mohamed Guerroumi[1],
and Mohamed Belaoued[3,4]

[1] Faculty of Computer Science, USTHB University, Bab Ezzouar, Algiers, Algeria
r_bitit@esi.dz
[2] Center of Excellence in Information Assurance (CoEIA), King Saud University,
Riyadh, Saudi Arabia
abderhab@ksu.edu.sa
[3] CReSTIC EA 3804, University of Reims Champagne Ardenne, Reims, France
mohamed.belaoued@univ-reims.fr, m.belaoued@caplogy.com
[4] Caplogy, Poissy, France

Abstract. Cybersecurity incidents have dramatically increased in the recent years. The number of cyber-attack variants is also increasing, thus traditional signature and behavioral-based intrusion detection techniques cannot adapt to continuously changing cyber-attack patterns. In addition, tradilçtional intrusion detection approaches are reactive and respond to a specific behavior or signature, and detect cyber-attack after its occurrence. Therefore, in order to effectively deal with cyber-attack, there is a need for new approaches allowing to act proactively against cyber-attacks by forecasting the future cybersecurity incidents and alert system administrator. For this purpose, we design a new approach which is composed of three main components: data gathering, multivariate forecasting, and Fuzzy Inference-based Decision System (FIDS). Data gathering extracts the multivariate time series of network traffic, and feeds them into multivariate forecasting-component which transforms it into a supervised learning dataset. This dataset allows a machine learning model to predict future cyber-attacks rate and relative features. The FIDS component uses the predicted variable of multivariate times series, i.e. cyber-attack rate and relevant features, to evaluate the risk of cyber-attack at the next time step. The proposed approach was evaluated using CICDDoS2019 dataset, and the results show that Support Vector Regression has the best forecasting accuracy compared to other models.

Keywords: Cyber-attack forecasting · Predictive model · Fuzzy Inference System · Machine learning · Multivariate time series

1 Introduction

With the growing usage of Internet and information technologies, accessing and sharing information becomes increasingly easy. However, the usage of these

E. Renault and P. Mühlethaler (Eds.): MLN 2022, LNCS 13767, pp. 24–35, 2023.
https://doi.org/10.1007/978-3-031-36183-8_3

technologies is not without risk of cyber-attacks towards individual's and organization's information systems.

Cyber-attacks are rapidly evolving in number and sophistication. On the other hand, defense approaches cannot keep up with the pace of cyber-attacks. Today's defense methods can be classified into reactive and proactive approaches. Reactive approaches aim to prevent and detect cyber-attacks by monitoring system information in real time, and attempt to identify any suspicious behavior. These approaches are based on attack patterns and signatures, and can only detect attacks that are already launched [21]. Otherwise, proactive approaches aim to act against cyber-attacks in advance, by analyzing network event history and predicting upcoming threat. Most of these approaches are based on statistical models, which deal with a time series forecasting. However, they are characterized by a univariate time series limited to cyber-attack rate [4].

In this paper, we propose a proactive method to overcome cyber-attacks, which is composed of three main components: data gathering component, multivariate forecasting component, and Fuzzy Inference-based Decision System (FIDS). To the best of our knowledge, this paper is the first to attempt to predict the rate of cyber-attacks using multivariate time series as supervised machine learning problem. In addition, it investigates the application of Fuzzy Inference System by utilizing forecasting results in the decision system. The main contributions of this paper are:

- We use a multivariate time series, which composed of cyber-attack rate and the most relevant features, and apply a correlation-based methodology to select the best relevant features to be used as variables for the multivariate time series.
- We use different machine learning models to forecast cyber-attack based on a multivariate time series.
- We design a Fuzzy Inference-based Detection System (FIDS), which is used to evaluate the risk level of the attack at the next time step.
- We evaluate the proposed solution using a recent DDoS dataset, named CICDDoS2019 dataset [2].

The rest of this paper is organized as follows: Sect. 2 presents related work. Section 3 gives a brief background on time series and Fuzzy Inference System. Section 4 describes our approach. Section 5 presents and discusses the implementation and evaluation results. Finally, the paper is concluded in Sect. 6.

2   Related Work

To defend against cyber-attacks in a proactive way, many approaches have been proposed, most of them are based on time series forecasting.

Sokol et al. [17] applied the Box-Jenkins methodology [1] to forecast future cyber-attacks using time series analysis. Two prediction approaches were used and compared: Auto-regressive and Bootstrap. The proposed solutions were evaluated using a time series data collected from the honeynet located in the CZ-NIC network [6].

In [20], Werner et al. used ARIMA model to predict the number of cyber-attacks for the next day based on historical attack count. The forecasting model was built using a dataset collected from the Hackmageddon [11] database.

In [12], the authors used ARMA and GARMA models to forecast future cyber-attack rates. These forecasting models were fitted using a dataset gathered from a Honeynet. The study shows that GARMA model performs better than ARMA model.

Silva et al. [15] proposed a method called One Point Analysis (OPA) to forecast the brute force attacks. The method uses Pseudo-Random Binary Sequences (PRBS), and Exponential Weighted Moving Averages (EWMA) to predict beginning of the burst.

In [3], the authors designed a method to forecast the impact of DDoS attack, rate (packets/sec) and size (number of used machines/bots). The model processes the data extracted from a darknet traffic, then it uses different forecasting models, namely, moving average, weighted moving average, exponential smoothing and linear regression.

Zhan et al. [22] used a combination of time series and the extreme value theory for predicting cyber-attack rate. This combination can lead to more accurate results.

Okutan et al. [10] used data that are extracted from Twitter social media and the open source GDELT [5] project, and fed them into a Bayesian classifier to predict future attacks against organizations.

In [8], the authors applied machine-learning techniques to predict future cyber-security incidents. The method uses a collection of commonly used IP address-based/host reputation blacklists and a set of security incident reports to train a support vector machine (SVM) for cyber-attack prediction.

Despite the presented results, the above related work methods suffer from the following limitations:

- Most of them use statistical models, which deal with univariate time series of cyber-attack rate.
- Some methods use more than one feature to forecast cyber-attacks, but they do not apply a methodology for feature selection when a large volume of raw data is available.
- The related work methods provide forecasting of future cyber-attack, but they do not explain the next step to do after forecasting, i.e., risk assessment or decision making.

Therefore, we address the earlier-mentioned limitations by:

- Using a multivariate time series, which composed of cyber-attack rate and the most relevant features.
- Applying a correlation-based methodology to select the best relevant features, which can be used as variables of the multivariate time series.

- Using different machine learning models to forecast cyber-attack based on a multivariate time series.
- Designing a Fuzzy Inference-based Detection System (FIDS), which is used to evaluate the risk level at the next time step.

3 Background

3.1 Multi-variate Time Series

A time series is a sequence of repeated observations of the same phenomenon on different dates, for example the average daily temperature in a given place, the number of cyber-attacks per day against a given organization, the average electricity consumption per month, etc. The values of time series are on the same scale indexed by a time parameter [19]. Time series forecasts are applied in a wide range of fields, including weather forecasting, economic forecasting, prediction of computer network traffic, etc.

Two types of time series can be distinguished, univariate and multivariate. Univariate time series is a series with a single time-dependent variable. Multivariate time series has more than one time-dependent variable, each variable does not only depend on its past values, it also has some dependence with other variables. It corresponds to the simultaneous observation of several time series. Multivariate time series highlights the effects of correlation and causality between different variables, which is used to predict future values.

3.2 Fuzzy Inference System

Fuzzy inference is a method that uses fuzzy logic [9] to map a given input to an output. It is based on a set of fuzzy rules in the form of IF/THEN. The notion of fuzzy rule [9] makes it possible to define a fuzzy expert system as an extension of a classical expert system, manipulating the fuzzy proposition.

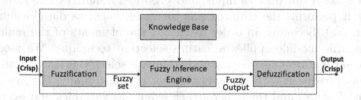

Fig. 1. Architecture of Fuzzy Inference System.

A fuzzy Inference System (FIS) is formed of three blocks [13], as shown in Fig. 1. The first one, named fuzzification block, transforms the numerical values into degrees of membership of the different fuzzy sets. The second block is the inference engine, which consists of a set of rules. Finally, the defuzzification block allows to infer a net value that results from the aggregation of rules.

4 Proposed Approach

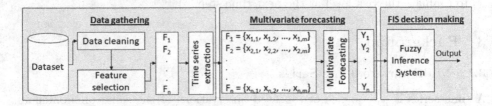

Fig. 2. Architecture of the proposed approach.

One of the limitations of related work is to apply forecasting using a univariate time series (i.e.. cyber-attack rate) and statistical models. Our approach aims at forecasting cyber-attack rate using multivariate time series and machine learning models. More precisely, we attempt to answer the following questions:

- What are the main characteristics that can be used as multivariate time series members to forecast cyber-attack rate?
- How can machine learning models perform compared to statistical models?
- How can we exploit the output of forecasting to make a decision and alert security administrators?

To answer thee above questions, we propose the approach shown in Fig. 2, which consists of three main components, detailed as follows:

4.1 Data Gathering

The main objective of this model is the selection of features and the generation of a multivariate time series, which is used to train and test the forecasting module. It takes a raw data as input, and produces a multivariate time series.

Firstly, it performs the required preprocessing (such as data cleaning, normalization, ...). Secondly, in order to choose the elements of the multivariate time series, this module applies a feature selection technique. We used a correlation based feature selection method, then we select the features which has the highest correlation with the cyber-attack events. Finally, the module uses the selected features and the cyber-attack events as variables and extracts the multivariate time series using a fixed time step.

4.2 Multivariate Forecasting

This component is used to make forecasting. It uses the past values of the multivariate time series to predict the values of the next time step.

This component takes as input the multivariate time series, and applies a forecasting models. For the machine learning models, we need to reframe the

$$F_1 = \{x_{1.1}, x_{1.2}, ..., x_{1.m}\}$$
$$F_2 = \{x_{2.1}, x_{2.2}, ..., x_{2.m}\}$$
$$\cdot$$
$$\cdot$$
$$\cdot$$
$$F_n = \{x_{n.1}, x_{n.2}, ..., x_{n.m}\}$$

Fig. 3. Frame of multivariate time series.

time series as supervised learning problems, so that the input of the model become a pairs of values, which represent the past values, and output, which represent the values at next time step, as illustrated in Fig. 4.

Therefore, the series must be transformed from its original form, as shown in Fig. 3, to the form of a supervised learning dataset, as shown in Fig. 4, where:

- F_i ($i = 1, \ldots, n$) is a variable of time series, which are selected from the original dataset.
- n is the number of the variable of time series including the cyber-attack rate.
- t is the number of the past values used to predict the next time step values.
- $x_{i,j}$ ($i = 1, \ldots, n$ and $j = 1, \ldots, t$) is the value of the variable F_i at the time step j.
- $x_{i,j}$ represent the input of the machine learning models.
- $x_{i,t+1}$ is the target output to be predicted.

By using this reframed time series of Fig. 4, the supervised machine learning model can learn how to predict the output value (a future value) from the input (past values). The reframed set will be used to train and test the machine learning models.

input										output		
$F_{1,t}$			$F_{2,t}$			...	$F_{n,t}$			$F_{1,t+1}$	$F_{2,t+1}$	... $F_{n,t+1}$
$x_{1.1}$	$x_{1.2}$	... $x_{1.t}$	$x_{2.1}$	$x_{2.2}$	... $x_{2.t}$	...	$x_{n.1}$	$x_{n.2}$	... $x_{n.t}$	$x_{1.t+1}$	$x_{2.t+1}$	...$x_{n.t+1}$

Fig. 4. frame of multivariate time series as supervised learning.

4.3 Fuzzy Inference-based Detection System (FIDS)

The FIDS component is used as decision support tool. It takes as input the forecasted values to predict the risk level of an attack at the next time step. First, it normalizes the predicted variables of cyber-attack rate and relevantfeatures, then it performs the FIS process, which is described in Fig. 1. The output of FIDS is an integer value comprised between 0 and 1, where 0 refers to the weakest risk level and 1 refers to the highest risk level.

The linguistic variables of FIDS are the variables of time series, i.e. cyber-attack rate and selected features, whereas the corresponding characteristics {ʼvery highʼ, ʼhighʼ, ʼmediumʼ, . . .} are known as the linguistic terms [9]. In our case, the output linguistic variable is the risk, which is characterized by the linguistic terms ʼhighʼ, ʼmediumʼ, and ʼlowʼ.

5   Implementation and Evaluation Results

In this section, we present the implementation steps and the experimental results. To implement the proposed approach, we used Python programming language with scikit-learn [14], scikit-fuzzy [16] and statsmodels [18] packages to implement the different components and test various models using CICDDoS2019 dataset [2]. We used the mean squared error (MSE) as an evaluation metric. The MSE is calculated using Eq. 1 [7]:

$$MSE = T^{-1} \sum_{t=1}^{T} (y_t - \widehat{y}_{t|t-1})^2 \tag{1}$$

5.1   Data Gathering and Forecasting Results

By applying the preprocessing and feature selection on SYN attacks of the dataset, we found out that the most relevant features related to the cyber-attack rate are the two features *"Inbound"* and *"ACK Flag Count"*, which represent the amount of inbound traffic and number of ACK packets respectively.

Figure 5 and Fig. 6 show the evolution of cyber-attack flows as function of ʼInboundʼ and ʼACK Flag Countʼ respectively. High correlation is observed between these features and cyber-attack, which means that these features have a high impact on the prediction of the attack. Therefore, we use the cyber-attack rate, ʼInboundʼ, and ʼACK Flag Countʼ as variables of the multivariate time series. Then, the multivariate time series is reframed according to the process described in Sect. 4.2. The obtained dataframes were used to train and test the following models: Random Forest Regression, Support Vector Regression, Linear Regression, LASSO Regression, and Vector Autoregression. The latter is a statistical model, which is designed for multivariate forecasting. Table 1 shows the MSE for each model, and for each variable of the time series, as well as the mean MSE per model. It shows an average MSE between 213 and 470 for machine learning models, and 855 for the statistical model Vector Autoregression. These results show that machine learning models have more ability to

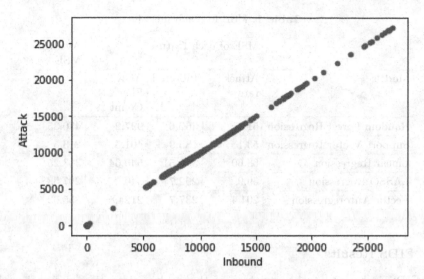

Fig. 5. Attack flows as function of 'Inbound'.

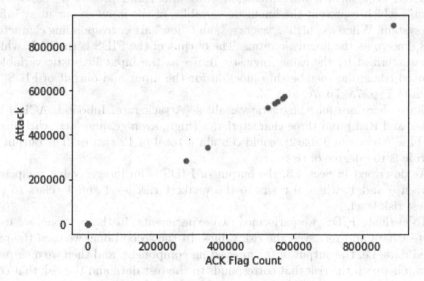

Fig. 6. Attack flows as function of 'ACK Flag Count'.

predict the future from historical data than the Vector Autoregression model. By comparing the machine learning models, we can observe that Support Vector Regression has better performance than other models with an average MSE of 213.5, followed by LASSO Regression and Linear Regression. Random Forest Regression has the lowest performance in terms of average MSE, despite it has the best forecasting accuracy of cyber-attack rate. This is due to the high MSE values for the features 'Inbound' and 'ACK Flag Count'.

Table 1. MSE for each model

Models	MSE of each feature			Average MSE
	Attack rate	Inbound	ACK Flag Count	
Random Forest Regression	51.2	362.6	997.2	470.35
Support Vector Regression	53.1	185.9	401.5	213.5
Linear Regression	68.60	300.51	540.64	303.25
LASSO Regression	59.9	292.7	540	297.53
Vector Autoregression	204.4	237.7	2124.8	855.63

5.2 FIDS Results

The FIDS is based on the variables of the multivariate time series, i.e. the predicted cyber-attack rate, predicted "Inbound", and predicted "ACK Flag Count", which represent the linguistic variables of the input of our fuzzy inference system. Whereas "high", "average", and "low" are corresponding characteristics, known as the linguistic terms. The output of the FIDS is the Risk, which is characterized by the same linguistic terms as the input linguistic variables. We used triangular membership function for the input and output of FIDS, as shown in Figs. 7a, 7b, 7c, and 7d.

Since there are four linguistic variables (Attack rate, Inbound, ACK Flag Count, and Risk) and three characteristics (high, average, and low), each fuzzy rule base (shown in Table 2), could contain a total of 12 entries. The output of the rule is the degree of risk.

As described in Sect. 4.3, the output of FIDS is an integer value comprised between 0 and 1, where 0 refers to the weakest risk level and 1 refers to the highest risk level.

To evaluate FIDS, we performed two experiments. In the first one, we used real test data and retrieved the risk values. In the second one, we used the predicted data, i.e. the output of the forecasting component, and then we measured the gap between the risk that corresponds to the test data and the risk that corresponds to the forecast data using the MSE. Table 3 shows the MSE of FIDS for each case.

We can observe that the decisions made based on predicted data of Support Vector Regression model have the best accuracy compared to that of the other models. In addition, we can notice a clear superiority of the decisions based on predictions made by the machine learning models compared to Vector Autoregression.

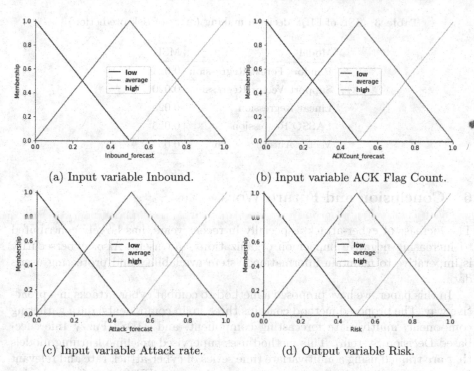

(a) Input variable Inbound.

(b) Input variable ACK Flag Count.

(c) Input variable Attack rate.

(d) Output variable Risk.

Fig. 7. Membership functions of the inputs and output of FIS decision-making.

Table 2. Rule base of FIDS

Attack rate	Inbound	ACK Flag Count	Risk
high	high	high	high
high	high	average	high
high	average	low	high
high	average	high	high
average	low	average	average
average	low	low	average
average	high	high	high
average	high	average	high
low	average	low	average
low	average	high	average
low	low	average	low
low	low	low	low

Table 3. MSE of FIDS decision making (error of risk prediction)

Model	MSE
Random Forest Regression	0.024
Support Vector Regression	0.020
Linear Regression	0.029
LASSO Regression	0.025
Vector Autoregression	0.053

6 Conclusion and Future Work

The increase of cyber-attack, especially in recent years, has largely contributed to increasing negative impact on organizations and users. Also, cybersecurity is imperative to maintain information system availability and protecting user's data.

In this paper, we have proposed a method to combat cyber-attacks in a proactive way. The proposed method contains three main components: data gathering component, multivariate forecasting component, and FIDS (Fuzzy Inference-based Decision System). This method uses supervised machine learning models that are trained using a multivariate time series of cyber-attack rate and relevant features to predict cyber-attack rate at next time step. FIDS uses the prediction result to evaluate the future risk at the next time step by applying the Fuzzy Inference System concept. The proposed approach was evaluated using CICD-DoS2019 dataset. Experimentation results show that Support Vector Regression model has the best forecasting accuracy compared to other models.

This work introduces a proactive approach to combat cyber-attacks by monitoring past events and forecasting cyber-attack rate and the risk level of next time step. In next work, we would extend our system to perform forecasting and risk evaluation for multi-step attack scenarios. In addition, this work can be improve by using modern deep learning models, and optimizing the rule base of FIDS.

Acknowledgement. The authors would like to thank Caplogy for supporting this work, which is the result of their collaboration in 2022.

References

1. Box, G.E., Jenkins, G.M., Reinsel, G.C., Ljung, G.M.: Time Series Analysis: Forecasting and Control. Wiley, Hoboken (2015)
2. Canadian institute for cybersecurity: DDoS evaluation dataset (CICDDoS2019) (2019). https://www.unb.ca/cic/datasets/ddos-2019.html. Accessed 22 Sep 2022
3. Fachkha, C., Bou-Harb, E., Debbabi, M.: Towards a forecasting model for distributed denial of service activities. In: 2013 IEEE 12th International Symposium on Network Computing and Applications, pp. 110–117. IEEE (2013)

4. Gandotra, E., Bansal, D., Sofat, S.: Computational techniques for predicting cyber threats. In: Jain, L.C., Patnaik, S., Ichalkaranje, N. (eds.) Intelligent Computing, Communication and Devices. AISC, vol. 308, pp. 247–253. Springer, New Delhi (2015). https://doi.org/10.1007/978-81-322-2012-1_26
5. The GDELT project. http://www.gdeltproject.org. Accessed 20 Aug 2022
6. The honeynet project. https://www.honeynet.org. Accessed 20 Aug 2022
7. Hyndman, R.J., Athanasopoulos, G.: Forecasting: Principles and Practice. OTexts, Heathmont (2018)
8. Liu, Y., Zhang, J., Sarabi, A., Liu, M., Karir, M., Bailey, M.: Predicting cyber security incidents using feature-based characterization of network-level malicious activities. In: Proceedings of the 2015 ACM International Workshop on International Workshop on Security and Privacy Analytics, pp. 3–9 (2015)
9. Nedjah, N., de Macedo Mourelle, L.: Fuzzy Systems Engineering: Theory and Practice, vol. 181. Springer, Berlin (2005). https://doi.org/10.1007/b102051
10. Okutan, A., Yang, S.J., McConky, K.: Predicting cyber attacks with bayesian networks using unconventional signals. In: Proceedings of the 12th Annual Conference on Cyber and Information Security Research, pp. 1–4 (2017)
11. Passeri, P.: Hackmageddon: information security timelines and statistics. http://www.hackmageddon.com. Accessed 20 Aug 2022
12. Pillai, T.R., Palaniappan, S., Abdullah, A., Imran, H.M.: Predictive modeling for intrusions in communication systems using GARMA and ARMA models. In: 2015 5th National Symposium on Information Technology: Towards New Smart World (NSITNSW), pp. 1–6. IEEE (2015)
13. Sabri, N., Aljunid, S., Salim, M., Badlishah, R., Kamaruddin, R., Malek, M.: Fuzzy inference system: short review and design. Int. Rev. Autom. Control 6(4), 441–449 (2013)
14. scikit-learn. https://scikit-learn.org/stable/. Accessed 17 Oct 2022
15. Silva, A., Pontes, E., Zhou, F., Guelf, A., Kofuji, S.: PRBS/EWMA based model for predicting burst attacks (brute froce, dos) in computer networks. In: Ninth International Conference on Digital Information Management (ICDIM 2014), pp. 194–200. IEEE (2014)
16. skfuzzy 0.2 docs. https://pythonhosted.org/scikit-fuzzy/. Accessed 17 Oct 2022
17. Sokol, P., Gajdoš, A.: Prediction of attacks against honeynet based on time series modeling. In: Silhavy, R., Silhavy, P., Prokopova, Z. (eds.) CoMeSySo 2017. AISC, vol. 662, pp. 360–371. Springer, Cham (2018). https://doi.org/10.1007/978-3-319-67621-0_33
18. statsmodels. https://www.statsmodels.org/stable/index.html. Accessed 20 Aug 2022
19. Watson, M.: Time series: economic forecasting. Int. Encycl. Soc. Behav. Sci. 20, 15721–15724 (2001). ISBN: 0-08-043076-7
20. Werner, G., Yang, S., McConky, K.: Time series forecasting of cyber attack intensity. In: Proceedings of the 12th Annual Conference on Cyber and Information Security Research, pp. 1–3 (2017)
21. Yang, S.J., Du, H., Holsopple, J., Sudit, M.: Attack projection. In: Kott, A., Wang, C., Erbacher, R.F. (eds.) Cyber Defense and Situational Awareness. AIS, vol. 62, pp. 239–261. Springer, Cham (2014). https://doi.org/10.1007/978-3-319-11391-3_12
22. Zhan, Z., Xu, M., Xu, S.: Predicting cyber attack rates with extreme values. IEEE Trans. Inf. Forensics Secur. 10(8), 1666–1677 (2015)

iPerfOPS: A Tool for Machine Learning-Based Optimization Through Protocol Selection

Hamidreza Anvari$^{(\boxtimes)}$ and Paul Lu

Deptartment of Computing Science, University of Alberta, Edmonton, AB, Canada
{hanvari,paullu}@ualberta.ca

Abstract. Previous work established the importance of selecting the right network protocol for new foreground traffic, based on the current background traffic. The interactions between protocols, such as TCP-CUBIC and TCP-BBR for congestion control, affect fairness and throughput on shared networks. Fortunately, machine-learned (ML) classifiers can be used to identify the current background protocols, then optimization through protocol selection (OPS) can be used to improve performance on shared wide-area networks (WAN).

We describe the design, implementation, and evaluation of iPerfOPS, the first tool that uses OPS to perform bulk-data transfer. The new tool is a substantially modified version of the well-known iPerf tool, and is an end-to-end implementation that incorporates previous research results. iPerfOPS introduces (1) a reliable data-transfer capability to iPerf, and (2) an implementation of OPS. We describe some empirical evaluations of iPerfOPS and discuss some of the practical implementation details required to achieve high performance. iPerfOPS shows that it is possible, within one tool, to classify the background network protocols such that high throughput and fairness are achieved.

Keywords: iPerf · protocol selection · machine-learned classifier · data transfer · wide-area networks · fairness · shared network

1 Introduction

iPerfOPS is a new tool for bulk-data transfer on shared wide-area networks (WAN). Although based on the well-known iPerf measurement tool [1], a noteworthy element of iPerfOPS is its use of optimization through protocol selection (OPS) [10]. In previous work, the need for an appropriate network protocol selection was established. Empirical studies showed how suboptimal protocol selection can lead to low throughput and the unfair sharing of networks. With iPerfOPS, we attempt to bridge the gap from theory to practice.

© The Author(s), under exclusive license to Springer Nature Switzerland AG 2023
É. Renault and P. Mühlethaler (Eds.): MLN 2022, LNCS 13767, pp. 36–55, 2023.
https://doi.org/10.1007/978-3-031-36183-8_4

iPerfOPS is the first implementation of an end-to-end tool to (Fig. 1) (1) collect round-trip time (RTT) data, (2) perform OPS via machine-learned (ML) classifiers, (3) change the congestion control algorithm (CCA) used by the foreground data transfer, and (4) show high throughput. The performance experiments in this paper evaluate the new tool and the practical aspects of making it work well.

In theory, it is known that if a background data stream is already using TCP with the BBR (Version 1)[1] CCA (TCP-BBR), then the new foreground data stream should also use TCP-BBR (instead of, say, TCP-CUBIC) because BBR is known to be unfair to CUBIC data streams [7,17]. In contrast, two TCP-BBR data streams tend to be fair to each other. And, conversely, if a background data stream is already using TCP-CUBIC, then the new foreground data stream should also use TCP-CUBIC, or else the foreground transfer might be unfair to the background transfer. Recognizing the protocols in use by background streams before selecting the protocols for the new foreground stream can be important for maintaining high performance and fairness. In theory, ML classifiers can detect the CCA in current use by the background traffic [7,8].

In practice, we now show that iPerfOPS can use ML classifiers to select the appropriate CCA for the foreground. We demonstrate that the ML classification works even if the CCAs in the background traffic change during the data transfer.

Our contributions include:

1. We introduce iPerfOPS, the first tool to use OPS for bulk-data file transfer. The original iPerf measures network performance, but does not actually implement reliable *file* transfer.
2. We document key design decisions required to make iPerfOPS into a high-performance file-transfer tool.
   (a) Architectural changes are required for iPerf (e.g., socket pools, Sect. 4.2).
   (b) A new active probing pattern (Sect. 5) is required for OPS to work if the foreground stream has already been on the network.
3. We discuss a performance evaluation of iPerfOPS, showing that the new tool achieves fair network throughput, even in the presence of a dynamically changing background CCA protocol.

2 Background and Related Work

2.1 Dedicated Vs. Shared Networks: Impact of Background Traffic

Networks are usually either *dedicated* or *shared*. In dedicated networks, a private data path is established to connect several points of presence, usually for large-scale and data-intensive research projects or industrial use. For example, Google's B4 private world-wide network [19], and Microsoft's private WAN

---

[1] While BBR version 2 has been under development, at the time of writing this paper, BBR version 1 is still the only stable version publicly available; hence the one used in this study for all the experiments and evaluations.

resources [23] are dedicated networks. Bandwidth reservation techniques also exist to allocate dedicated bandwidth for a window of time [15].

In contrast, bandwidth-sharing networks are still the common case for a large number of academic and industrial users. On a shared network, the available bandwidth is shared among all the users and applications in a competitive environment. Bandwidth sharing can create a dynamic workload on the network, with consequences such as periodic burstiness in the traffic mix [18,20]. This burstiness may result in various levels of contention on the network, increasing the packet loss rate. As a result, bandwidth utilization can be negatively impacted, both in aggregate and for individual users.

On dedicated networks, the goal for data-transfer protocols is maximum bandwidth utilization. For example, GridFTP [6] and UDT [3], two well-known high-performance tools, are mostly concerned with end-to-end throughput. In contrast, on shared networks, the fair sharing of bandwidth with other network users (i.e., fairness) is important [5].

2.2 TCP Scheme: Congestion Control Algorithm (CCA)

TCP is the most popular reliable transport protocol on the contemporary Internet, as well as on many private networks. In addition to reliability features, TCP includes a *congestion control algorithm (CCA)* that dynamically adjusts the sending rate to optimize bandwidth utilization, while maintaining cross-stream fairness [5].

CUBIC [16] and BBR [12] are two popular state-of-the-art CCAs. CUBIC is the default CCA on most Linux machines, and BBR is a recent algorithm from Google, that has been gaining popularity in recent years. While most CCAs include fairness as part of their design goals, fairness problems have been reported for both CUBIC and BBR [22,24,31]. Several solutions have been proposed by the research community to address issues with BBR [26,28,30].

Another recent trend in CCAs is to apply ML techniques to the design or optimization of protocols. Remy uses off-line simulation to create a new TCP CCA using a Markov decision process [29]. Performance-oriented Congestion Control (PCC) uses an on-line learning approach for its TCP CCA [13]. Vivace [14] and Proteus [25] are two newer CCAs based on PCC, trying to improve on its utility functions.

2.3 OPS: Optimization Through Protocol Selection

As discussed earlier, background traffic streams on the shared network impact various performance metrics (e.g., bandwidth, latency) of a data-transfer task. Hence, obtaining knowledge about the background traffic would allow for optimizations to network configurations (e.g., choosing appropriate data-transfer tools and protocols).

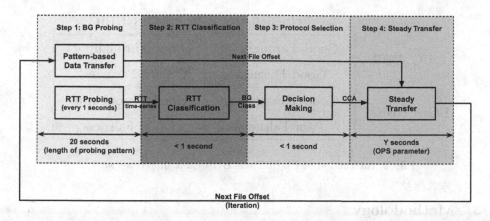

Fig. 1. Iterative OPS operation cycle, comprised of four main steps: (1) Background Probing (active/passive), (2) RTT Classification, (3) Protocol Selection, (4) Steady Transfer.

In one previous study, we devised a method to classify the mixture of background traffic, recognizing different mixtures of TCP-CUBIC and TCP-BBR on the network with an accuracy of up to 85% [7]. In that approach, we periodically probe end-to-end delay at regular time intervals (e.g., every 1 s) to generate a time-series representation of RTT over time. We refer to this probing method as *passive probing*. We then use the resulting RTT time series for training a classifier to identify the mixture of TCP-based traffic on the network. That knowledge of the background traffic is central to optimization through protocol selection (OPS) [10] for a variety of network bandwidths and latencies [9].

We extended the *passive probing* (RTT only) technique, by introducing an *active probing* [8] element to the model. By perturbing the network traffic with short bursts of traffic, the resulting RTT signatures are better for classification, improving accuracy to (up to) 95%. The intuition behind active probing is that different network protocols have distinct, transitory reactions (i.e., not steady state) to a competing traffic stream.

In the OPS strategy, depicted in Fig. 1, we iteratively probe the RTT (Step 1), classify the RTT time-series (Step 2), and pick a protocol, TCP-CUBIC or TCP-BBR (Step 3), to transfer data (Step 4). If the background CCAs change, the next iteration of this operation cycle will detect the change and (re-)select the foreground CCA (Step 3) to optimize throughput and fairness. For the rest of this study, we will use this operation cycle and the classifiers from our previous studies, in the new end-to-end data transfer tool, iPerfOPS.

	FG: BBR	FG: CUBIC
BG: BBR	Good Throughput Good Fairness	Poor Throughput Poor Fairness
BG: CUBIC	Good Throughput Poor Fairness	Good Throughput Good Fairness

Fig. 2. Interaction of CUBIC and BBR on a shared network

3   Methodology

In this section, we review our research methodology, including the ML setup as well as the network testbed and utilized software configurations.

3.1   ML Process

Scope: TCP-CUBIC Vs. TCP-BBR. While a diverse collection of tools and protocols are in use on the Internet, TCP-based protocols currently dominate traffic on data networks. Hence, in this study we only consider TCP-based background traffic. To keep this work comparable with our previous work, we limit our study to mixes of TCP-CUBIC and TCP-BBR as the background traffic. According to previous work [7,9], the interaction between TCP-CUBIC and TCP-BBR can be qualitatively summarized as in Fig. 2.

In particular, we will consider six distinct classes of background traffic to train a classifier for them. These six classes represent various mixtures of up to two streams of CUBIC and BBR, as summarized in Fig. 3c.

Classification Task. As discussed in Sect. 2.3, we follow the same ML process as in our previous studies to train the classification model [7,8]. In particular, we use a 1-Nearest Neighbor (1-NN) classifier trained using RTT time-series data from active probing, with length $w = 20$.

Time-series classification is a special case where the input data being classified are time-series vectors. While Euclidean distance is widely used as the standard distance measure in many ML applications, it is not well-suited for time-series data, since there is noise and distortion over the time axis [21]. More sophisticated approaches are needed for comparing time-series data. We utilize the dynamic time warping (DTW) distance measure along with 1-NN for classifying the RTT time-series data. DTW has a warping parameter for which we tested different values, settling at 10.

Active Probing Patterns. As discussed earlier, we conduct active probing in the form of short bursts of traffic. To trigger background streams to reveal their unique patterns, as opposed to a single traffic burst, we use a sequence of short traffic bursts with different configurations.

We have scripted our experimental environment with the probing profile as a module. This would enable us to extend our experiments to more probing profiles in the future. For example, one *active probing pattern* could be defined as a sequence of three segments of traffic bursts (CUBIC for 5 s, BBR for 5 s, CUBIC for 5 s), followed by a silence period of 5 s. We use the abbreviation *C5-B5-C5-S5* for referring to this pattern. In the process of generating data for training, as well as when deploying the classifiers, the active probing pattern is conducted repeatedly while RTT is being probed.

3.2 Network Testbed

For the testbed, we employ the same emulated network as our previous studies, depicted in Fig. 3a [4,7]. The emulated network is implemented as an overlay on top of a physical network. All the nodes in our testbed are physically located in a dedicated computer cluster. There is no interference from other traffic because the cluster is isolated from other networks. The hardware configuration of the nodes are provided in Fig. 3b. The nodes are equipped with dual network interface cards of 1 Gb/s and 10 Gb/s native rates. In this paper, we only present results conducted on the 1 Gb/s network.

The emulated WAN link is throttled to 500 Mb/s bandwidth and 65 ms latency. Furthermore, the router buffer size at nodes R1 and R2 are set to 6 MB, corresponding to 1.5×BDP. This buffer size should avoid the effects of a shallow buffer (<<BDP) and the extended delays of bufferbloat caused by deep buffers (>>BDP).

3.3 Libraries and Software Configuration

We used the following software and frameworks:

1. **OS Kernel.** All the nodes run CentOS 6.4 Linux, using kernel version `4.12.9-1.el6.elrepo. x86-64`.
2. **TCP CCAs.** We use two CCA schemes in this study: CUBIC and BBR. For CUBIC, we use the default version shipped with the OS kernel. As mentioned earlier, BBR Version 1 is the only stable version publicly available, hence we use BBR v1 for all the experiments in this study.
3. **Network customization.** We use the Dummynet network emulator [11]. Our automation scripts for data-gathering and evaluation are implemented in either Python or Bash shell scripts.
4. **Machine Learning Environment.** All the ML programming and evaluation metrics are implemented in Python. For implementing K-NN classifiers, we use the scikit-learn library version 0.19.1 [27].

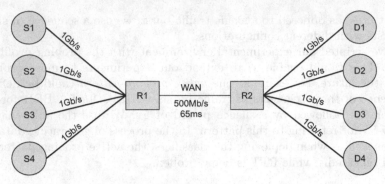

(a) Dumbbell Network Topology

Node(s)	CPU (Model/Cores/Freq.)	RAM
S1, D1	AMD Opt. 6134 / 8 / 2.30	32 GB
S2, D2	Intel Ci3-6100U / 4 / 2.30	32 GB
S3, D3	Intel Ci3-6100U / 4 / 2.30	32 GB
S4, D4	AMD E2-1800 / 2 / 1.70	8 GB
R1, R2	AMD A8-5545M / 4 / 1.7	8 GB

Traffic Class	Num. BBR	Num. CUBIC
B0-C0	0	0
B0-C1	0	1
B1-C0	1	0
B1-C1	1	1
B1-B1	2	0
C1-C1	0	2

(b) **Nodes Configuration**   (c) **Background Traffic Classes**

Fig. 3. Testbed Architecture and Experiment Configuration

4   iPerfOPS: An OPS Data-Transfer Tool

We implemented an end-to-end OPS-based data-transfer tool, named iPerfOPS. As its name suggests, iPerfOPS is based on the well-known *iPerf* [1] network benchmarking tool, adding new functionality for data-transfer intervals, interleaved with periodic probing of the background traffic. There is no existing, network-probing feature in iPerf, nor is iPerf designed to reliably bulk-data transfer the contents of a file. Yet OPS requires periods of network probing to generate the input features to the ML classifier. Therefore, substantial modifications were required to iPerf.

Nonetheless, we adopt iPerf because it is a well-known, trusted benchmarking tool and it is highly configurable for fixed-time and fixed-size traffic tests. Moreover, iPerf is a stable, well-optimized, open-source tool, with great portability across different platforms.

We present the high-level design and implementation of iPerfOPS. First, we provide an in-depth discussion on adopting iPerf as the underlying data-transfer tool and the extensions we added to iPerf to support OPS use cases. Second, we provide the high-level architecture of iPerfOPS as a standalone, end-to-end, data-transfer tool. In the next section, we provide end-to-end performance results for iPerfOPS.

4.1 Development Strategy: Adopting iPerf Library

In our efforts to implement OPS as an end-to-end, production-ready, data-transfer tool we considered two options: First, build OPS from scratch, using the native Linux interfaces and libraries (e.g., network sockets, files). Second, leverage an existing tool with a programmable API or library, and reuse its high-level data-transfer constructs and abstractions. While the first option provides finer control for building underlying constructs and workflows from scratch, the second option empowers us to take benefit from a wide range of well-designed and stable community-provided tools. The use of existing tools could also positively contribute to the overall quality of our resulting tool, while saving us on required development efforts.

For the second option, building on an existing tool, there are two main categories: data-transfer tools, and network benchmarking tools.

Since OPS is designed to improve data-transfer efficiency, the data-transfer tools seem to be the best fit for this purpose. TCP-based tools, such as GridFTP [6], are immediately capable of serving as the data-transfer channel of a given TCP CCA. Furthermore, UDP-based data-transfer tools, such as UDT [3], provide flexibility in fine tuning and customization of CCAs to be used for data transfer. Meanwhile, the lack of two important features in data-transfer tools renders them insufficient for building OPS.

First, data-transfer tools are incapable of conducting fixed-time transfers. They are mainly built for transferring a full file to the completion. While they may be tweaked, by the means of I/O redirects or otherwise, to partially transfer a file, they will still function based on the byte-count and size of file (i.e., fixed-size). However, the OPS strategy requires the capability of transferring data blocks in a fixed-time fashion to allow conducting different patterns for active-probing, and to transfer data for a fixed period of time before re-probing the background traffic.

Second, data-transfer tools usually function as black-box tools, merely providing any insights or metrics of the data-transfer performance. In most cases, the end-to-end throughput, reported directly or via the proxy of transfer time, is the only performance metric available. Those tools usually use various internal metrics and counters to optimize the data transfer, but such metrics remain internal to the tool and are not accessible, or meaningful, to be used as performance proxies to the outside world.

Network benchmarking tools, in contrast to data-transfer tools, mostly provide flexibility in terms of end condition (fixed-time and fixed-size). In addition, those tools by design provide a larger range of performance metrics. Some of the well-known tools in this group include Netperf [2], and iPerf [1]. The available performance metrics based on the tool in use could include average throughput, instantaneous (interval) throughput, jitter, latency, and more. The presence of described properties makes this group of tools a desirable candidate to be utilized for implementing the OPS.

iPerf is a network benchmarking tools with an active community and on-going development and maintenance. In addition, iPerf serves its full functionality via an API library, making it a great networking software for building new tools. Lastly, iPerf open-source software, enabling us to make any necessary adjustment to its source-code. Hence, we decided on using iPerf as the underlying library for implementing OPS. We should note that iPerf has a legacy version known as iPerf2, which is now replaced with iPerf3. Throughput this chapter any references to iPerf means iPerf3, unless otherwise mentioned.

For OPS, iPerf has two limitations, one functional and one performance-related. First, the functional inadequacy, is the lack of reliable file transfer. iPerf supports reading data from a file for probing the network. However, it does not implement reliability in file transfer, and terminates the data transfer socket as soon as the termination criteria is reached on the sender (i.e., specified byte-count or seconds-length). Second, the performance inadequacy, is that iPerf creates and destroys network sockets per invocation, making it I/O-intensive if there are multiple subsequent invocations happening (e.g., in active probing patterns for OPS [8]). In the following section we will discuss and report on our approach on modifying and extending iPerf to address these inadequacies.

4.2   Extending iPerf for Data Transfer

As discussed, iPerf required two modifications to accommodate our OPS implementation. First, iPerf lacks reliable file-transfer functionality. Second, iPerf re-establishes socket connections frequently, which required some optimization.

iPerf Architecture and Operation State-Machine. The iPerf design follows a client-server model, operating based on a stateful application-level protocol and using a distinct control channel and data channel(s) for client/server coordination. The iPerf state-machine is also provided in Fig. 4.

iPerf client and iPerf server use a control channel for three distinct use cases:

1. Initial test setup: swapping test parameters and initiating the task.
2. Ongoing sync: during the test client and server use control socket to inform and prompt the other party to transition to a new state.
3. Communicating results: at the end of the data transfer, client and server swap their statistics before terminating the task.

First, the iPerf is started on in server mode on destination node. iPerf server by default binds and listens to port 5001 (which may be customized). Once the server is up, the other nodes may run iPerf in client mode taking the server IP for traffic conduction. Then, iPerf client starts by communicating the server instance. After an initial hand-shaking between the client and the server, data channel(s) are created and the requested benchmarking takes place which involves a fixed-time or fixed-size data transfer using the provided configuration. A few of relevant iPerf configurations include the following:

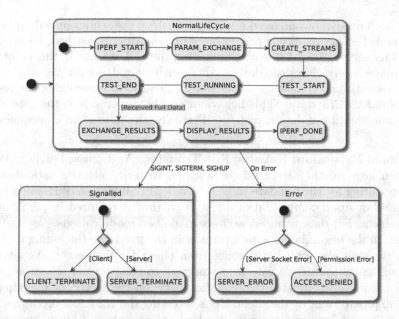

Fig. 4. iPerf State Machine. Both client and server follow the same state transitions. The state transitions are mostly coordinated by the server, except for TEST_END and IPERF_DONE, or if client is signalled (CLIENT_TERMINATE).

1. Transport protocol: it supports TCP, UDP, and SCTP. Default is TCP.
2. Number of data streams: it supports single or parallel data streams for data transfer. Default is a single stream.
3. CCA: in case of TCP and SCTP transports, the CCA could be selected. The available options and default CCA are the same as the host running iPerf.
4. Data direction: the data transfer might be directed from client to server, or vice versa. Default is from client to server.
5. Data source: the data on the sender could be randomly generated on the fly, or a file may be used to read data from. Default is random data.
6. Data destination: the received data on the receiver could be discarded, or a file may be specified for writing the received data. Default is discarding the data.
7. End condition: a byte count (fixed-size) or the time length in seconds (fixed-time) may be specified as end condition. Default is fixed-time for 10 s.

The iPerf server is designed as a single-tasked service, meaning that it could only engage in a single data transfer at any given time. As soon as a client request arrives and the connection is accepted as a control channel, the server becomes unavailable and any new clients trying to connect while the current server engagement is not completed will get denied with an error message. Once the current experiment is completed, both client and server communicate results and then terminate all sockets and cleanup. Server then goes back to listen for new incoming connection requests.

For both fixed-time and fixed-size data transfers, the sender puts data on the data channel socket(s). When the termination condition is reached, client sends a signal for server (TEST_END state) indicating the end of test. At this point, the server immediately closes the data sockets (without waiting for the whole data to be received), and signals the client that it's ready for exchanging the results (EXCHANGE_RESULTS state). This behaviour on the server side is the core cause for the unreliability of data transfer in iPerf (i.e., the functional inadequacy).

Functional Extension: Reliable File Transfer. As discussed earlier, iPerf's operation logic on the server side is designed to terminate the data channels without waiting for the full data to be received. Terminating a connection based on a timer or upon completion of a test is all that is required in performance measurement. For data transfer, we have made two modifications.

First, in the beginning of a data transfer in the fixed-size (including fixed-file mode), we send the total bytes count from client for the server. As a result, server will know the size of data it is expecting to receive from the client.

Second, we have adjusted the operation cycle on the server for handling the TEST_END prompt sent by the client. Upon receiving the TEST_END prompt, server enters a timeout loop and periodically checks the number of bytes successfully received on the data channel. It delays closing data channel until the full data is successfully received. Server, then, terminates the data channel and sends EXCHANGE_RESULTS prompt for the client.

Furthermore, in order to accommodate partially transfer data from a file on the client through different intervals (as with the OPS probing patterns), we added a new functionality to iPerf, allowing the sender to partially send data from a file by seeking to a location and sending specified bytes count. We also adjusted server to add a mode to append the received data to the end an existing file, rather than replacing the existing file as implemented in the original iPerf.

Performance Optimization: Pooling Network Sockets. The periodic probing interval in the OPS and the idea for constructive active probing implies a sequence of short data-transfer episodes until the data transfer is completed. As such, the efficiency of creating those short data-transfer episodes is a critical quality for achieving production-ready performance.

However, as we already discussed in Sect. 4.2, iPerf creates, configures, and destroys control and data sockets per each transfer, which makes it inefficiently resource-intensive for our use-case. One common pattern to predictably control the overhead in applications with high frequencies of socket use is to establish a *Socket Pool*. In this pattern, a special data structure takes the ownership and accounts for creating, maintaining, and reusing network sockets. When our application requires a socket it will request the socket pool. If an open and unused socket is available to the desired destination it will be allocated for the use. Otherwise a new socket is created. When we are done with the socket, it will be returned to the socket pool and marked as available for any subsequent request for a socket.

While a socket pool seems to be the right fit to optimize the socket use in iPerf, there exists a challenge in adopting this pattern with iPerf: iPerf concurrently uses control and data socket(s) between a client-server pair of nodes, following significantly different semantics for each of control and data sockets. So we need to coordinate client and server on which existing socket to be used for control or data channel for a new episode of data transfer. Furthermore, due to the relatively complex state-machine in iPerf and the highly formalized use of control socket for communicating the state transitions, we ideally should not disrupt the client-server control protocol as it might make it incompatible with the original iPerf library for the future updates.

In order to overcome the challenge of coordinated socket reuse, we have devised a *Typed* Socket Pool, where each socket will be identified as either Control or Data socket. Since we exactly need one of each socket type for any data transfer, it would then suffice to keep track of one instance of each socket type to be reused through the lifespan of an OPS invocation. In addition, to minimize the refactoring efforts and divergence from original iPerf, we intend to keep this socket pooling and reuse mostly transparent to the iPerf modules and functions.

In our ultimate solution we have created wrapper functions for the basic socket library functions. Each wrapper function named and signature is identical to its corresponding library function, but named with Capital case initial letter. For the two special functions that create a new socket, `socket()` and `accept()` we also created typed wrapper functions (`TSocket()` and `TAccept()`) which take an additional parameter *SocketType*. The types methods help the wrapper functions to return the existing socket of the desired type when requested, which enables the wrapper functions to remain agnostic about the iPerf internal logic of control versus data channels. We also implemented a data structure `SocketRegistry` for pooling typed sockets. It is worth noting that the SocketRegistery is a simplified version of socket pooling for the special case of iPerf library where we only need one of each socket type at a time, and we only have a single data-transfer task at any given time.

4.3   iPerfOPS Architecture

iPerfOPS follows a client-server model. The high-level modular design of iPerfOPS is provided in Fig. 5. The operation cycle for iPerfOPS is also provided in Fig. 6. Upon invocation, iPerfOPS starts an infinite loop of data-transfer episodes until the end of file is reached. Each data-transfer episode consists of two phases, Probing phase and Steady phase. The probing phase follows passive or active probing policy [8] based on the provided command-line parameters. The result of probing is then used to classify the background traffic and selecting a CCA (e.g., CUBIC or BBR) to be used for the subsequent steady phase. The steady phase uses the provided CCA for transferring data from the file for a configured time length of Y seconds (Fig. 6. The longer the steady state, the fewer and longer transfer episodes it will take for a file transfer, which in turn implies less frequent probing phases. We will further discuss the impact of steady phase length in the Sect. 5.

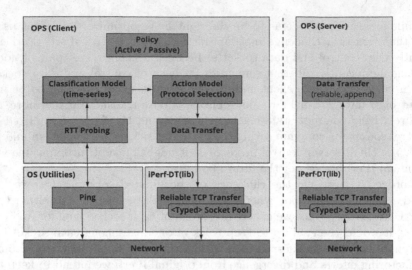

Fig. 5. iPerfOPS Client/Server Architecture

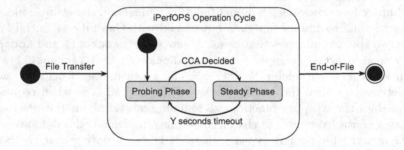

Fig. 6. iPerfOPS operation Cycle (handled at the client side)

iPerfOPS Server. The server relies on iPerf to accept incoming connections, and to reliably write the received data to the user-specified file. It operates to append-mode, always appending data to the end of the target file. Server design follows a simpler logic compared to the client. It is run by providing a port number as well as the file path for storing the received data. The server remains agnostic about the state of the data transfer whether it is in probing phase or steady phase of data transfer. The client handles the logic for different episodes of data transfer.

iPerfOPS Client. The iPerfOPS client orchestrates the episodes of data transfer, interleaved steady phases of data transfer with periodic probing phases. It relies on the *ping* utility provided by the operating system for probing RTT and generating RTT time-series. The RTT time-series are then passed to the classification module. The classification results are then passed to Action module

where a decision is made on the CCA to be used for the following steady phase
of data transfer.

While our current implementation contains a minimal viable version of functionality, the modular design of iPerfOPS allows us to independently extend and
improve each module independently in the future. For example, the classification
module, which now operates based on the 1-NN with DTW (Sect. 3.1), may be
extended in the future to use more sophisticated classic or deep time-series classification models. Similarly, the action module which currently contains a simple
logic may be extended with more advanced decision making models for picking
a transfer protocol, considering more parameters such as network configuration,
policies, and more.

5   Performance Evaluation

In this section we discuss data-transfer experiments and report on the end-to-
end performance results. We use the same networking testbed as presented in
Sect. 3.2.

Background Traffic. In our previous OPS evaluations, we have studied
the OPS performance in the presence of a constant background pattern, either
CUBIC, BBR, or a mixed of both. To further study the adaptive end-to-end
performance of iPerfOPS, we add new test scenarios where the background traffic
alternates between CUBIC and CCA in regular time intervals. We use ALT-X to
identify those configurations, where X represent the time interval in seconds that
one CCA runs before toggling the CCA.

All the reported results in this chapter present data-transfer tasks, transferring 14 GB data files. The average throughput over 5 runs are provided, and the
error bars present standard deviation.

Probing Patterns. While both passive and active probing policies are supported by iPerfOPS, in this section we only present and discuss constructive
active probing policy as it outperforms passive probing policy in terms of higher
classification accuracy [8]. So far we have introduced the C5-B5-C5-S5 as the
only active probing pattern available during our evaluations. However, in order
to better study and identify the importance and capability of distinct probing
patterns to drive classification and decision making, in this section we add a
second probing pattern to better identify the impact of silence periods in the
probing patterns. The new probing pattern is S5-C5-S5-B5, where we have 5 s
silence period before sending 5 s of CUBIC or BBR traffic as part of the probing
pattern. With this new pattern we aim to better identify the RTT pattern of
background traffic without our probing interference.

Results: The Impact of Distinct Probing Pattern. Fig. 7, especially part
(b), shows the throughput improvement gained by using iPerfOPS. In (b), all
of the (dynamic) OPS-[100,360] bars have much higher throughput (all between
210 and 305 Mb/s) than the CUBIC bar (about 50 Mb/s). Given the 500 Mb/s

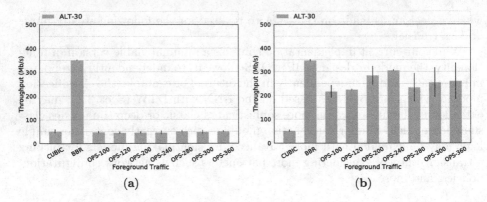

Fig. 7. iPerfOPS Performance in presence of alternating background traffic. (a) Active probing pattern: C5-B5-C5-S5 (non-effective) (b) Active probing pattern: S5-C5-S5-B5 (effective). All results are average over 5 runs, transferring 14GB random file. The results correspond to ALT-30 background pattern (single constant TCP stream of background traffic, toggling between CUBIC and CCA every 30 s). OPS-Y represents foreground traffic of probing network followed by sending data steady for Y seconds and repeat.

bottleneck bandwidth, the 250 Mb/s level represents a fair share of the network. At 50 Mb/s, the CUBIC bar represents much less than a fair share of the bandwidth, and is the result of a static choice of CUBIC for the foreground CCA.

The BBR bar (almost 350 Mb/s) shows how unfair BBR is against CUBIC, in which the ALT-30 background traffic alternates between CUBIC and CCA every 30 s.

Figure 7(b) shows that OPS can detect, classify, and then select an appropriate CCA for the foreground traffic that is effective in maximizing throughput. Note that the OPS task is non-trivial because the background traffic dynamically changes CCA (between CUBIC and CCA) every 30 s. Also, an important practical, implementation detail is that a S5-C5-S5-B5 active probing pattern (part (b))is required for OPS to be effective.

In contrast, the probing pattern C5-B5-C5-S5 (Figs. 7(a)) is ineffective in improving performance, presenting virtually identical performance as fixed CUBIC, the probign pattern S5-C5-S5-B5 (Figs. 7b) is effectively improving the data-transfer performance by several orders of magnitude compared to fixed CUBIC CCA, while treating the cross traffic more fairly compared to the aggressive BBR.

The distinct results between the two different active probing patterns shows the effect, and the importance, of the probing pattern in use by the OPS for probing phase. The probing pattern C5-B5-C5-S5, which was effective in improving performance of our proof of concept experiments in our previous study [7], proves ineffective in our end-to-end evaluation of iPerfOPS. In contrast, the pattern S5-C5-S5-B5 improves the ability of generated RTT time-series in identifyingbackground mixture, hence improved throughput performance compared to CUBIC and better fairness compared to BBR. We may attribute, among other reasons, the improved performance for S5-C5-S5-B5 to the added silence periods prior to conducting traffic using different CCAs. This enables the generated RTT time-series to better manifest the unique patterns for different types of background traffic.

Furthermore, increasing the steady state length, improves the performance for the effective pattern S5-C5-S5-B5 (Fig. 7b). The longer steady state periods mean less silence periods in total, hence better throughput performance. However, there is a trade-off when setting the steady state length for OPS. The shorter the steady state, the more frequent probing phases will be, and OPS sooner gets the opportunity to identify the changes in the background traffic patterns and adjust the CCA accordingly. In contrast, the longer the steady state, the less time is spend on probing phase, hence less time in silence periods and overhead of segment-wise active probing patterns. Hence the performance could improve if the background traffic is not changing too often.

One general rule of thumb is to set steady state to longer values if the frequency of the change in background traffic is low. For example on private networks or institution bandwidths where the pattern and type of bandwidth usage is more predictable. However, for public shared bandwidths or networks with a larger number of users the shorter steady state could help OPS to better adapt to frequent changes to the background traffic.

To further study the effect of the frequency of alternating background CCA, we have conducted the same experiment for different configurations for X in ALT-X background traffic pattern. The results for X values of 30, 60, 120, and 300 are provided in Fig. 8. The throughput performance is calculated in presence of fixed and alternating background traffic. All results are average over 5 runs, transferring 14GB data files containing random binary data. ALT-X represents background traffic of alternating CUBIC and BBR every X seconds. OPS-Y represents foreground traffic of probing network followed by sending data steady for Y seconds and repeat.

Figure 8(a) presents results in probing pattern C5-B5-C5-S5 in use, while Fig. 8(b) presents results for probing pattern S5-C5-S5-B5. While there are throughput variations based on the standard deviation (error bars), the overall performance trend remains the same across different configurations of ALT-X for the background traffic.

Once again (as with Fig. 7), Fig. 8(b) shows that OPS is effective in achieving fair throughput against dynamically changing background traffic, this time with greater interval variability. And, once again, the probing pattern of S5-C5-S5-B5 is key to enabling OPS to properly detect, classify, and perform OPS.

While in current OPS design the steady state length is specified as a user-provided configuration, one possible improvement to for the future is to auto-adjust the steady state length. So when OPS identifies changes in background traffic in subsequent probings, it should decrease the steady state length to better capture dynamic patterns of background traffic. But if subsequent probing

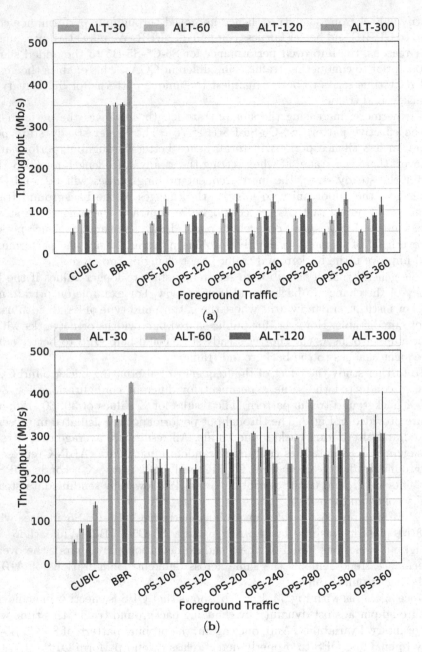

Fig. 8. iPerfOPS Performance in presence of fixed and alternating background traffic. (a) Active probing pattern: C5-B5-C5-S5 (non-effective) (b) Active probing pattern: S5-C5-S5-B5 (effective). All results are average over 5 runs, transferring 14GB random file. ALT-X represents background traffic of alternating CUBIC and BBR every X seconds. OPS-Y represents foreground traffic of probing network followed by sending data steady for Y seconds and repeat.

phases result in the same background traffic class, this implies less dynamic background patterns and hence we should increase the steady state length.

6   Concluding Remarks

We have discussed the design, implementation, and evaluation of iPerfOPS, an implementation of the OPS strategy. We show that iPerfOPS provides a tangible performance improvement for end-to-end data transfer when configured with appropriate probing patterns. Our tool extends the functionality of the original iPerf with the OPS functionality, adding reliable and efficient data transfer.

While our empirical evaluation of iPerfOPS shows the effectiveness of OPS, it also reminds us of the importance of implementation details, such as the probing pattern. In theory, and in previous papers, the C5-B5-C5-S5 pattern should have been effective. In practice, the implementation and evaluation of iPerfOPS revealed that the S5-C5-S5-B5 pattern is required to make OPS work as an end-to-end tool.

For future work, the impact of timing considerations (e.g., the ability to gather accurate RTT time series data, with 1 s intervals, under all circumstances) and other practical factors should be studied. In fact, we speculate that the impact of the foreground traffic itself, which was not a factor in our previous work, is important for protocol classification during an real data transfer, as observed with the new probing pattern. As iPerfOPS is refined further, other gaps between theory and practice will likely emerge.

References

1. iperf: Bandwidth measurement tool. http://software.es.net/iperf/
2. Netperf network benchmarking. https://hewlettpackard.github.io/netperf/
3. Udt: Udp-based data transfer for high-speed wide area networks. Comput. Netw. **51**(7), 1777–1799 (2007), protocols for Fast, Long-Distance Networks
4. Anvari, H., Lu, P.: The impact of large-data transfers in shared wide-area networks: An empirical study. Procedia Comput. Sci. **108**, 1702–1711 (2017). International Conference on Computational Science, ICCS 2017, 12–14 June 2017, Zurich, Switzerland
5. Afanasyev, A., Tilley, N., Reiher, P., Kleinrock, L.: Host-to-host congestion control for TCP. IEEE Commun. Surv. Tutor. **12**(3), 304–342 (2010)
6. Allcock, W., et al.: The globus striped gridftp framework and server. In: Proceedings of the 2005 ACM/IEEE Conference on Supercomputing, pp. 54-. SC '05 (2005)
7. Anvari, H., Huard, J., Lu, P.: Machine-learned classifiers for protocol selection on a shared network. In: Renault, É., Mühlethaler, P., Boumerdassi, S. (eds.) MLN 2018. LNCS, vol. 11407, pp. 98–116. Springer, Cham (2019). https://doi.org/10.1007/978-3-030-19945-6_7
8. Anvari, H., Lu, P.: Active probing for improved machine-learned recognition of network traffic. In: Machine Learning for Networking, pp. 122–140 (2020)

9. Anvari, H., Lu, P.: Learning Mixed Traffic Signatures in Shared Networks. In: Krzhizhanovskaya, V.V., Závodszky, G., Lees, M.H., Dongarra, J.J., Sloot, P.M.A., Brissos, S., Teixeira, J. (eds.) ICCS 2020. LNCS, vol. 12137, pp. 524–537. Springer, Cham (2020). https://doi.org/10.1007/978-3-030-50371-0_39

10. Anvari, H., Lu, P.: Machine-learned recognition of network traffic for optimization through protocol selection. Computers **10**(6) (2021). https://doi.org/10.3390/computers10060076, https://www.mdpi.com/2073-431X/10/6/76

11. Carbone, M., Rizzo, L.: Dummynet revisited. SIGCOMM Comput. Commun. Rev. **40**(2), 12–20 (Apr 2010). http://doi.acm.org/10.1145/1764873.1764876

12. Cardwell, N., Cheng, Y., Gunn, C.S., Yeganeh, S.H., Jacobson, V.: Bbr: Congestion-based congestion control. Queue **14**(5), 50:20–50:53 (Oct 2016)

13. Dong, M., Li, Q., Zarchy, D., Godfrey, P.B., Schapira, M.: PCC: Re-architecting congestion control for consistent high performance. In: 12th USENIX Symposium on Networked Systems Design and Implementation (NSDI 15), pp. 395–408. Oakland, CA (2015)

14. Dong, M., et al.: PCC vivace: Online-learning congestion control. In: 15th USENIX Symposium on Networked Systems Design and Implementation (NSDI 18), pp. 343–356. Renton, WA (2018)

15. Guok, C., Robertson, D., Thompson, M., Lee, J., Tierney, B., Johnston, W.: Intra and interdomain circuit provisioning using the oscars reservation system. In: Broadband Communications, Networks and Systems, 2006. BROADNETS 2006 3rd International Conference on, pp. 1–8 (Oct 2006). https://doi.org/10.1109/BROADNETS.2006.4374316

16. Ha, S., Rhee, I., Xu, L.: Cubic: a new tcp-friendly high-speed tcp variant. SIGOPS Oper. Syst. Rev. **42**(5), 64–74 (2008)

17. Hock, M., Bless, R., Zitterbart, M.: Experimental evaluation of bbr congestion control. In: 2017 IEEE 25th International Conference on Network Protocols (ICNP), pp. 1–10 (2017). https://doi.org/10.1109/ICNP.2017.8117540

18. Hong, C., et al.: Achieving high utilization with software-driven wan. In: Proceedings of the ACM SIGCOMM 2013 Conference, pp. 15–26. SIGCOMM '13 (2013)

19. Jain, S., et al.: B4: Experience with a globally-deployed software defined wan. In: Proceedings of the ACM SIGCOMM 2013 Conference on SIGCOMM, pp. 3–14. SIGCOMM '13 (2013)

20. Jiang, H., Dovrolis, C.: Why is the internet traffic bursty in short time scales? In: Proceedings of the 2005 ACM SIGMETRICS Intl. In: Conference on Measurement and Modeling of Computer Systems, pp. 241–252. SIGMETRICS '05 (2005)

21. Keogh, E., Ratanamahatana, C.A.: Exact indexing of dynamic time warping. Knowl. Inf. Syst. **7**(3), 358–386 (2005)

22. Kozu, T., Akiyama, Y., Yamaguchi, S.: Improving rtt fairness on cubic tcp. In: 2013 First International Symposium on Computing and Networking, pp. 162–167 (Dec 2013). https://doi.org/10.1109/CANDAR.2013.30

23. Liu, H.H., et al.: Efficiently delivering online services over integrated infrastructure. In: 13th USENIX Symposium on Networked Systems Design and Implementation (NSDI 16), pp. 77–90. Santa Clara, CA (2016)

24. Ma, S., Jiang, J., Wang, W., Li, B.: Towards RTT fairness of congestion-based congestion control. CoRR abs/1706.09115 (2017). http://arxiv.org/abs/1706.09115

25. Meng, T., Schiff, N.R., Godfrey, P.B., Schapira, M.: PCC proteus: Scavenger Transport and Beyond. ACM, New York, NY, USA (2020)

26. Pan, W., Tan, H., Li, X., Li, X.: Improved Rtt fairness of Bbr congestion control algorithm based on adaptive congestion window. Electronics **10**(5), 615 (2021). https://doi.org/10.3390/electronics10050615

27. Pedregosa, F., et al.: Scikit-learn: machine learning in Python. J. Mach. Learn. Res. **12**, 2825–2830 (2011)

28. Song, Y.J., Kim, G.H., Cho, Y.Z.: BBR-CWS: Improving the inter-protocol fairness of bbr. Electronics **9**(5) (2020)

29. Winstein, K., Balakrishnan, H.: TCP ex machina: Computer-generated congestion control. In: Proceedings of the ACM SIGCOMM 2013 Conference on SIGCOMM, pp. 123–134. SIGCOMM '13 (2013)

30. Yang, F., Wu, Q., Li, Z., Liu, Y., Pau, G., Xie, G.: Bbrv2+: Towards balancing aggressiveness and fairness with delay-based bandwidth probing. Comput. Netw. 206, 108789 (2022). https://doi.org/10.1016/j.comnet.2022.108789

31. Zhang, Y., Cui, L., Tso, F.P.: Modest BBR: Enabling better fairness for BBR congestion control. In: 2018 IEEE Symposium on Computers and Communications (ISCC), pp. 00646–00651 (2018). https://doi.org/10.1109/ISCC.2018.8538521

GRAPHSEC – Advancing the Application of AI/ML to Network Security Through Graph Neural Networks

Pedro Casas[1(✉)], Juan Vanerio[2], Johanna Ullrich[2,3], Mislav Findrik[4], and Pere Barlet-Ros[5]

[1] Austrian Institute of Technology, Vienna, Austria
pedro.casas@ait.ac.at
[2] University of Vienna, Vienna, Austria
juan.vanerio@univie.ac.at
[3] SBA Research, Vienna, Austria
jullrich@sba-research.org
[4] Cyan Security, Vienna, Austria
mislav.findrik@cyansecurity.com
[5] Universitat Politècnica de Catalunya, Barcelona, Spain
pere.barlet@upc.edu

Abstract. The application of Artificial Intelligence (AI) and Machine Learning (ML) to network security (AI4SEC) is paramount against cybercrime. While AI/ML is today mainstream in domains such as computer vision and speech recognition, it has produced below-par results in AI4SEC. Solutions do not properly generalize, are ineffective in real deployments, and are vulnerable to adversarial attacks. A fundamental limitation is the lack of AI/ML technology specific to network security. Network security data is intrinsically relational, and graph-structured data representations and Graph Neural Networks (GNNs) have the potential to drastically advance the AI4SEC domain. In this positioning paper we propose GRAPHSEC, a research agenda to systematically integrate GNNs in AI4SEC. We structure the state of the art in AI4SEC and on the application of GNNs to network security applications, elaborate on the benefits and challenges faced by GRAPHSEC, and propose a research agenda to advance the AI4SEC domain through GNNs.

Keywords: Network Security · AI4SEC · Graph Neural Networks

1 AI4SEC and the Present State of Affairs

Cybercrime is proliferating at an ever-growing rate and occurring everywhere on the Internet. Just like the evolution of technology, cybercrime must also morph to survive. This is why cyber criminals are constantly creating new attack types to fit and exploit new trends, while at the same time tweaking existing attacks to avoid detection. The early realization that traditional cybersecurity approaches

É. Renault and P. Mühlethaler (Eds.): MLN 2022, LNCS 13767, pp. 56–71, 2023.
https://doi.org/10.1007/978-3-031-36183-8_5

are failing to keep pace with the constant surge of new cyber attacks has opened the door to a relatively new research field which we shall refer-to as AI4SEC - the application of Artificial Intelligence (AI) and Machine Learning (ML) to cybersecurity. The complexity of cyber attacks, the ever-growing attack surface, the large amounts of high-dimensional data, and the disproportionate impact of attackers as compared to defenses, make of AI4SEC a paramount technology against cybercrime. However, while AI/ML is today mainstream in domains such as Computer Vision and Speech Recognition (CVSR), traditional AI/ML has produced below-par results in AI4SEC, and making of AI4SEC an accepted and fruitful basis in the cybersecurity practice has proven extremely challenging. Solutions do not properly generalize, are ineffective in real deployments, and are vulnerable to adversarial attacks. **A fundamental limitation is the lack of AI/ML technology specific to network security**.

1.1   AI4SEC – A Slow(ed) Path to Success

The impressive success of AI/ML, and in particular of Deep Learning (DL) [1], in multiple data-driven problems over the past decade has revamped the AI4SEC field; still, making of AI/ML an accepted and fruitful basis to cybersecurity in practice has proven extremely challenging. While research in AI4SEC has been very active for several years now, research outcomes on improving network security applications through AI/ML have had a limited impact on production systems to date. The mismatch is more notorious when considering the success enjoyed by other domains where AI/ML has brought tremendous innovation.

A common trend found in the existing literature is that, for the most part of the papers doing AI/ML for network security, there is a systematic lack of analysis on the multiple aspects which could lead to eventually re-use and reproduce, generalize, or even apply the obtained results in real deployments. In the seminal work of Vern Paxson [14], he elaborates on the challenges and limitations faced by AI/ML for network security when applied into real world deployments. Unfortunately, even if this paper dates back to 2010, there was never a follow up research agenda building up on top of these challenges to improve even some of the flagged limitations. Some of the most relevant challenges faced by AI4SEC to reach its full potential and adoption include:

① **Data complexity:** the Internet is a complex tangle of networks, protocols, technologies, applications, services, devices, and users. The interaction among all these components makes of the resulting data a major challenge when it comes to learning out of it.

② **Data diversity:** even within a single network, the network's most basic characteristics - e.g., the mix of different applications, can exhibit immense variability, rendering them unpredictable over short time intervals.

③ **Data dynamics:** networking data is dynamic by nature, and it is full of constant concept drifts - changes in the underlying statistical properties, which requires different from traditional approaches to make sense and good use out of it.

④ **Lack of ground truth:** supervised learning needs massive amounts of labeled data to learn, but "in the wild" networking data is mostly non-labeled; for the aforementioned challenges, labeling data in operational environments is extremely costly and error prone.

⑤ **Lack of learning generalization:** because of the data complexity, diversity, and dynamics, it becomes extremely difficult in the practice to learn models which can generalize to other environments, different from those where the training data comes from.

⑥ **Highly imbalanced tasks:** (network) security tasks are by nature complex to handle from a data-balance perspective, as anomalies and attacks reasonably occur much less often than normal-operation traffic.

⑦ **Lack of benchmarks:** different from other AI/ML-driven domains, where well established, publicly available datasets are available for testing, evaluation and benchmarking purposes (e.g., ImageNet in image processing), it is challenging to find appropriate public datasets to assess AI4SEC.

⑧ **Lack of performance bounds & high cost of errors:** deploying AI-based solutions in operational environments comes with the associated challenge of incurring in costly errors – especially when dealing with critical applications such as security, which usually ISPs and network vendors are not willing to bear. Being data-driven by nature, it is challenging to provide tight performance bounds on trained models.

⑨ **Learning occurs in an adversarial scenario:** model learning and execution occurs in an adversarial setting for security, where both the learning data and the trained models are prone to adversarial attacks, either by polluting the training process itself or by crafting the inputs such that the models are fooled.

⑩ **Lack of model transparency:** the lack of transparency of most AI/ML models limits their application in real deployments, especially in critical applications such as security. If you cannot understand why a certain model decision is taken, then you would not trust it and therefore not use it.

⑪ **Lack of combined expertise in AI/ML and network security:** the network security and AI/ML communities are disjoint, and it is hard to find the required expertise to properly tackle AI4SEC. There is a strong need for interdisciplinary approaches and teams to tackle this limitation.

The main hypothesis guiding GRAPHSEC's *raison d'être* is that AI/ML, if (and when) correctly applied to network security, could drastically improve the security of the Internet, and eventually the security of the society as a whole, reaching the levels of success enjoyed today by many other fields where AI/ML is strongly empowering innovation. But for this to happen, significantly more effort must be directed right into the core of the problem, as flagged by the aforementioned challenges.

1.2   AI4SEC and Graph Machine Learning

A lesson learned from the recent history of DL is that Neural Networks (NN) do not perform well when they are directly applied as a black-box to any type of problem and data. Instead, it is essential to carefully select and design models according to the nature of the data they will process, and introduce some biases that can help extract deep insights from the data seen during training, also known as inductive biases [4]. Indeed, from its inception the DL field has incrementally proposed new NN architectures to respond to the demands of different applications in different domains, with a main focus on the type of data they needed to process. For instance, the earliest fully-connected NNs showed weak performance when applied to CVSR problems (i.e., images and text). As a result, Convolutional NNs and Recurrent NNs emerged as new NN architectures particularly designed to extract meaningful information from images and sequences respectively.

From a geometric perspective, modeling the data structure results in a clear benefit, as it enables to learn and exploit fundamental symmetries and structural patterns from the target data [2], which is essential to then generalize to other data unseen during training (i.e., achieve combinatorial generalization [4]). Indeed, this combinatorial generalization mechanism reasonably resembles the way humans process, structure, and reason about the information received from the environment.

In this context, Graph Neural Networks (GNN) [3,4] have recently emerged as a new NN family particularly designed to learn and generalize over graph-structured data. Unlike previous NN architectures, GNNs present unique properties to abstract complex relational patterns between the different elements in graphs (i.e., they have a strong relational inductive bias [4]). Due to their unique ability to learn and generalize over graph-structured information, GNNs have recently enabled groundbreaking applications in multiple fields where data are generally represented as graphs (e.g., chemistry, physics, biology, social networks) [5]. GNNs are a family of deep learning methods based on neural networks, particularly suitable for the analysis of data described by graphs. These models implement specific mechanisms to exploit the relational information behind the data (e.g., they are equivariant to node and edge permutation).

Computer networks comprise relational information at many different levels (e.g., topology, user connections, flow interactions, knowledge graphs), which represent complex relationships and dependencies that are crucial for detecting many security threats; and graphs are precisely the most suitable mathematical abstraction to represent this relational information. For example, many network security attacks rely on complex multi-flow strategies (e.g., DDoS, network intrusions, port scans) that can be only unambiguously characterized by considering the structural relational patterns between flows. In general, to effectively detect this type of attacks, it is essential not only to capture relevant features on individual elements (e.g., flows, devices), but also the relationships between them (i.e., their inherent structural patterns). GNNs thus represent a well-suited DL method to process such information and abstract a more general knowledge – by exploiting the relational patterns in graphs.

To process graphs, GNNs initially build individual representations on graph elements, and dynamically build their internal NN architecture based on the elements and their connections in the graph (i.e., the graph edges). Then, the model executes a NN-driven message-passing algorithm, where elements iteratively exchange information on their neighborhood (i.e., with their connected elements). After propagating the information along the graph through message passing, each graph element eventually has an associated embedding that encodes relevant information about it and its local context in the graph, with the possibility to also produce embeddings at different levels of granularity (e.g., graph clusters, global embeddings). The main advantage of these novel NN models is that the generation of embeddings is completely data-driven, which benefits from the inherent capability of DL to process and learn from big amounts of data and, at the same time, incorporates for the first time a NN architecture that has the needed biases to learn complex relational patterns from graph-structured data (e.g., focus on relational reasoning [4], equivariance to node and edge permutation [2]). All these ingredients eventually allow GNNs to achieve proper combinatorial generalization and robustness over the target graph-structured information.

In the context of network security, an interesting aspect is the possibility to pre-train general-purpose GNN models (e.g., anomaly detection, data forecasting, unsupervised knowledge extraction), so that the embeddings generated by GNNs can be leveraged for different network security applications and use cases. Thus, more specialized DL models can be built on top of pre-trained GNN modules, by adding additional NN layers particularly designed for a target network security application, while they can still benefit from the ability of the underlying GNN module – via transfer learning – to capture complex relational patterns in the input data. As a result, these GNN-based models can offer the following main advantages with respect to state-of-the-art DL-based solutions: (i) better generalization and more robust characterization of attacks, (ii) more robustness to adversarial attacks, and (iii) better generalization to other networks and traffic.

1.3   The GRAPHSEC Vision – GNNs for Network Security

Network security data are intrinsically relational, and thus, we believe graph-structured representations and GNNs are foundational to AI4SEC, in the way convolutional and recursive networks were to CVSR. To advance on the application of GNNs and graph learning to network security application, we propose GRAPHSEC, a research agenda to systematically integrate GNNs in AI4SEC. The **goal of GRAPHSEC is to leverage graph data representations and modern GNN technology to conceive a new breed of robust GNN-based network security methods which could radically advance the AI4SEC practice**. The specific objectives of GRAPHSEC are: (a) to investigate algorithmic methods that facilitate modeling and learning from graph-based network security data; (b) to compare the benefits and overheads of GNN-based AI4SEC to traditional AI/ML in terms of detection performance, generalization, scalability, and robustness against adversarial attacks; (c) to showcase the benefits and improvements of GRAPHSEC technology in real-world network security applications as proof of concept.

The last few years have witnessed an increasing interest to exploit the potential of GNNs in networking applications [11,12], as many fundamental problems in networks involve graphs (e.g., topology, routing, protocol dependencies). For problems such as traffic engineering, GNNs have demonstrated their capability to effectively generalize to network topologies, traffic demands, and configurations unseen during training [13], representing a cornerstone in the conception of a new breed of ML-based solutions which can be successfully deployed in real-world networks. The AI4SEC domain has been traditionally more complex to tame [14], and as explained before, there are complex challenges and limitations faced by AI/ML for network security when applied to real-world deployments. GNNs could significantly advance many of these limitations, and the research agenda proposed by GRAPHSEC would unleash the power of GNNs in AI4SEC.

2 AI4SEC and GNN4SEC – State of the Art

Next, we overview the state of the art in the AI4SEC domain, in Graph Neural Networks, and on their application to network security applications. For the sake of completeness, and to position the state of the art on two of the most critical blocking points in AI4SEC, we also review related work on XAI (challenge (10)) and (synthetic) data generation for training (challenges (1) – (4)).

2.1 AI4SEC – AI/ML for Network Security

The application of AI/ML to cybersecurity has marked a fundamental shift in our ability to protect critical data systems and digital infrastructures. For strained security teams, it offers the possibility to keep pace with an ever-evolving threat landscape. While rule and signature-based solutions offer some protection against pre-identified threats, the reality is that attacks consistently evade these tools. Two different approaches are dominant in the research literature and commercial cybersecurity devices: signatures-based or misuse detection, and anomaly detection. Misuse detection is the de-facto approach used in most systems. When an attack is discovered, generally after its occurrence during a diagnosis phase, the associated malicious pattern is coded as a signature by human experts, which is then used to detect new occurrences of the same attack. Avoiding costly and time-consuming human intervention, signatures are also constructed by supervised AI/ML techniques, using instances of the discovered attack to build a detection model for it. Misuse detection systems are highly effective to detect attacks they are programmed to alert on. However, they cannot defend the systems against new attacks, simply because these attacks do not match their lists of signatures. Anomaly detection uses instances of normal-operation data to build normal-operation profiles, detecting anomalies as activities that deviate from this baseline. Such methods can detect new kinds of network attacks that were unseen before. Still, they require training to construct profiles, which is time-consuming and depends on the availability of anomaly-free traffic instances. In addition, it is not easy to maintain an accurate and up-to-date normal-operation profile, which induces high false-alarm rates.

Most approaches in the AI4SEC literature correspond to shallow learning, where the key relies on expert domain knowledge to carefully engineer the features used as an input to the model [18–20]. An interesting research line is the application of adaptive AI/ML-learning approaches to cybersecurity and anomaly detection. A relevant and representative example is presented in [15], where authors evaluate stream-based traffic-analysis approaches based on Hoeffding adaptive trees [16]. In [17] we explored the application of adaptive learning to the problem of adaptive defense against network attacks. Newer work addressing modern developments of AI/ML and their application to cybersecurity include results on deep learning [21–23], transfer learning [24], explainable AI for network security [25], adversarial learning [33], and more.

However, while general deep learning approaches are investigated in AI4SEC [47], the speed of adoption of AI4SEC solutions is slow, with significant reluctance on the applied domain. There is a striking gap between the extensive academic research and the actual deployments of such AI4SEC systems in operational environments, partially rooted in many of the *still unsolved* challenges faced by AI/ML for cybersecurity in the practice [14]. Even if the work from Paxson [14] dates back to 2010, there was never a comprehensive, follow up research agenda to improve the AI4SEC domain.

Summary: *despite the vast literature in AI4SEC, the exploitation of AI/ML in the cybersecurity practice requires new approaches and novel AI/ML technology to alleviate some of the core bottlenecks limiting their broader success, acceptability, and adoption.*

2.2 GNNs for NETSEC

GNNs represent a relatively new and very active research topic [3–9], due to their convincing performance in a plethora of domains. GNNs are neural networks that can be directly applied on graphs' data to tackle different analysis tasks at the node-level, edge-level, and graph-level, based on the concepts of *graph representation learning* or *embedding* to encode nodes, edges, or subgraphs' structural information into low-dimensional vectors [48]. There is a rich taxonomy of GNN models addressing multiple aspects of AI/ML [5]. Well-known variants of GNNs such as Graph Convolutional Network (GCN), Graph Attention Network (GAT), Graph Recurrent Network (GRN), GraphSAGE, have demonstrated ground-breaking performance on many deep learning tasks, and the study of different properties and limitations of GNNs is a very active area of research [8,50]. The design of graphs to improve learning performance [9,55], the generalization power of GNNs [6,7], their assessment and improvement in terms of robustness [51], the adoption of self supervision approaches [52], and the conception of models capable to learn from dynamic graph structures [53,54], are just some of the challenges under study today. **In the field of cybersecurity**, GNNs are gaining increasing attention as an effective approach to conceive more robust security threat detection methods. Recent papers have proposed the use of GNNs for malware detection in mobile applications [56,57], to exploit the relationship among network connections generated from different devices, showing a significant improvement with respect to the state of the art. Similarly, GNNs have been used for botnet detection [58], obtaining promising results. Our preliminary work on GNNs for network intrusion detection [59] has shown that we can conceive AI4SEC solutions which are significantly more robust against adversarial attacks than traditional ML-based methods.

Summary: *GNNs represent a very active research area, recently enabling groundbreaking applications, and driving the development of neural networks that are much more broadly applicable. While still on its infancy, the early work on the application of GNNs to AI4SEC applications suggest that graph-learning and GNNs could also represent a breakthrough for network security applications.*

2.3   XAI – Explainable AI for Network Security

While AI/ML technology offers unparalleled benefits when it comes to the analysis of complex systems and phenomena, their practical adoption is limited by the models' inability to explain decisions and actions to human users [26]. Explainable Artificial Intelligence (XAI) addresses this problem, and refers to AI/ML models yielding behaviors and predictions that are understandable by humans. They are important in application fields in which opaqueness, transparency and human reproducibility of machine learning predictions are of critical and practical importance. Cybersecurity is an example for such a field. In fact, the focus of most AI/ML research usually lies on prediction tasks and rarely on providing explanations/justifications on the model's behavior and decisions.

When it comes to the interpretability of results, traditional approaches are based on so-called white-box supervised techniques to explain the decisions provided by a particular AI/ML model. White-box techniques use simple and easy-to-interpret models such as linear discriminant functions or decision trees, which generally offer a straightforward interpretation of the models' decisions. However, white box explainability limits the set of applicable algorithms to those which are natively interpretable, with the additional drawback of potentially limiting performance. Recent work and efforts in XAI [27–29] provide both useful terminology and discussion on where and how such explainable approaches are useful. Current work has focused on black-box explanation methods, which require no knowledge about the model internals, and analyze it as a black-box through input/response analysis. Methodologies such as LIME [30] provide local interpretation methods to explain single model decisions/predictions by linearizing a general model around the specific inputs, identifying the most relevant input features for that prediction. This approach guarantees considerable flexibility in the model selection. Other current black-box approaches include SHAP [31], which provides global interpretation methods based on aggregations. XAI is mainly linked with supervised learning scenarios and relies on domain knowledge to interpret the proposed explanations. A recent XAI technique worth mentioning for the cybersecurity space is LEMNA [25], which extends LIME to improve local linearization. Other few studies have started focusing on the applicability of XAI concepts to the cybersecurity domain [32].

Summary: the XAI domain has gained significant relevance with the rapid growth of data and the increasing complexity of models, especially when it comes to assess the reliability of AI/ML systems for critical applications. However, in the cybersecurity domain, it is largely unclear how explainability of AI/ML and model transparency might be achieved in real-world applications.

2.4   Data Generation – Data Limitation and Lack of Labels

Datasets are crucial and their availability is, even today, among the biggest challenges for the application of AI/ML in cybersecurity. First and foremost, data sets are necessary for training of security systems [14,34]. Beyond, data sets are needed for their evaluation and comparison. Academia predominantly relies on publicly available data sets facilitating comparison of detection approaches [34]; however, these datasets bear drawbacks: (1) they might contain personal data, e.g., IP addresses, as defined by the General Data Protection Regulation (GDPR) and potentially also Intellectual Property (IP) when being recorded in real-world environments, effectively hindering their publication. (2) As a matter of fact, accessible data sets are typically limited in scope – otherwise maintenance, particularly labeling and anonymization, require high efforts - and lack representativeness of real networks. (3) Recency is also a matter of discussion with multiple datasets being older than 10 years. Research shows that the Internet is constantly evolving regarding traffic, latency, protocols, and services [35,36]. Sooner or later, even high-quality datasets become outdated and are inappropriate for training, evaluation, or comparison of contemporary security systems. In corporate environments like financial institutions or insurance companies, data handling in the context of data-driven, AI/ML-based security systems is debatable, too. The latter is on-time trained with (pre-)recorded traffic serving as a baseline. As retraining is a tedious and fragile task, it is barely done. Consequently, the baseline soon does not reflect the actual environment anymore. For supervised approaches, also the question on how to label traffic arises [37].

An alternative approach are traffic generators synthesizing data [37]. A taxonomy of synthetic traffic generators [38] includes (I) replay engines replaying previously recorded traffic in the same order and timing, (II) maximum throughput generators creating high amounts of traffic for a benchmark of network connections, (III) model-based generators relying on stochastic models for data generation, and (IV) high-level/auto-configurable generators measuring parameters of real data and synthesizing data accordingly. The authors emphasize that the individual validation processes of generators frequently lack quantitative assessments (e.g., comparison with actual data). Further, there is no consensus on validation parameters within the community. Both concerns are still valid [39] and again get to the point of limited representativeness and up-to-dateness of synthetic data.

Novel generative models such as those represented by Generative Adversarial Networks (GANs) [42], as well as more traditional approaches such as Variational

Auto-Encoders (VAEs) are powerful approaches to learn the underlying probabilistic distributions of data samples, in a purely data-driven, model-agnostic manner. Such models can be used in the practice to generate new synthetic samples [40,41], following the learned distributions from a given dataset. Using such powerful, data-driven synthetic generators allows for better and more complete means for doing so called data augmentation, namely enlarging the size of learning datasets by adding controlled variations to the original dataset, aiming at more robust and generalized learning. Data augmentation has been traditionally used in image processing [44], by creating modified versions of training images. GAN based generation additionally permits to control the trade-off between sample fidelity and variety [43], offering means to augment data in newly, unexplored directions. Finally, of relevance to the synthetic generation of data is the topic of automatic data labeling; while in principle it is easy to assign labels to synthetically generated data – after all, it is a controlled process, the usage of data-driven approaches might require techniques to assign labels to such new data in some robust way. Approaches such as label propagation and label spreading [45,46] are two popular graph-based methods, which propagate given labels from a (typically small) subset of the data samples to the whole dataset.

Summary: *the matter of data for (re-)training, evaluation, and comparison is key for the development of AI/ML-based security systems. At present, the potential of AI4SEC is not exploited to the full, both in academia and industry, due to data shortage.*

3 The Path to GRAPHSEC

3.1 Research Questions and Hypotheses

To integrate GNNs in AI4SEC, we propose to perform research on the main properties which would drastically advance the application of AI/ML to cybersecurity applications. For an AI4SEC application to be successful, the following criteria should be fulfilled: it must be **generalizable**, so that it can be applied and re-used in different deployments; it should operate with a low level of supervision in terms of ground-truth, seldom available in the cybersecurity domain – as **self-supervised** as possible; it should be **robust against adversarial attacks**, especially those AI-driven; it should be **scalable**, to be applied on the massive scale of data available in networks; and it must be capable to **deal with temporal dynamics and concept drifts** in the data. GRAPHSEC builds on the hypothesis that graphs and GNNs applied to AI4SEC problems are a turning point in the evolution of network security. To explore this potential, and to assess the benefits and limitations of GNNs in AI4SEC, we have defined a research agenda structured under the following research questions:

Ⓐ **Generalization** – How Good are GNNs to Generalize in AI4SEC? GNNs' expressiveness to capture different graph structures is paramount to achieve generalization [6]. Can graph representations and GNNs provide more general, invariant abstractions for more generalizable AI4SEC? How well can they predict labels, e.g., a graph property, for unseen graphs? Could we pre-train general-purpose GNN models (e.g., anomaly detection, forecasting, knowledge extraction), to leverage the generated embeddings for multiple AI4SEC applications, as done today in NLP or CVSR?

(B) **Robustness** – How much more Robust is GNN-driven AI4SEC? Small, unnoticeable perturbations of graph structure can catastrophically reduce the performance of popular GNNs [51]. How much more robust are GNN-based attack detectors against adversarial attacks as compared to shallow and deep-learning based detectors, using traditional spatial and/or temporal data representations? Following initial results in the literature [51], we would additionally investigate the conception of graph-tailored adversarial attacks, directly targeting the structural properties of the representation.

(C) **Self-Supervision** – How to Extend Self-Supervision to GNNs in AI4SEC? Learning by self-supervision has achieved promising performance on natural language and image learning tasks. Self-supervised learning (SSL) enables the training of deep models on unlabeled data, removing the need of excessive annotated labels. Within the context of SSL using GNNs [52], how could we leverage the graph structure of attacks as partial ground-truth for SSL, counterbalancing the lack of labeled data in AI4SEC? AI/ML-based models require abundant labeled data, which is seldom available in real deployments. SSL would enable learning from orders of magnitude more data.

(D) **Scalability** – How to Scale GNNs to Large-scale AI4SEC Applications? The irregular nature of graph data and the large graph sizes impose serious scalability issues for GNN training [5]. Can we quantify this limitation for the targeted AI4SEC applications? And can we improve it, through better graph representations? Sample-based training has shown promising results in this problem [10].

(E) **Dynamic Data** – How to Integrate Time in GNN-driven AI4SEC? Security data is of a dynamic, temporal nature, and thus, the structure of the underlying graphs is also dynamic (e.g., evolving features or connectivity over time). Which are the best approaches to learn from spatio-temporal dynamic graphs, for standard graph learning applications such as node classification, edge prediction, and graph clustering? While dynamic graphs is an almost untouched research area, results from inductive GNNs [49,53] and Temporal Graph Networks [54] make a strong baseline for further research.

3.2   Expected Results from GRAPHSEC

GRAPHSEC represents the first systematic and principled approach exploring the application of GNNs to network security applications, considering specific limitations identified in the application of AI/ML to cybersecurity problems, with the aim of significantly improving the AI4SEC domain. The answers to the proposed research questions sum up to the following GRAPHSEC objectives:

① **New AI4SEC Approaches:** investigating and conceiving GNN-based algorithmic approaches that leverage graph-represented network security data to improve current limitations in AI4SEC is the core objective proposed by GRAPHSEC.

② **Assessment of GNN4SEC:** realized improvements would be assessed by quantitatively comparing GNN-based AI4SEC to traditional AI/ML in terms of (i) detection performance, (ii) generalization of representation-learning and pre-trained models to unseen data distributions, as well as re-usability into similar tasks (domain adaptation), (iii) scalability to large data representations, and (iv) robustness against adversarial attacks.

③ **GRAPHSEC Prototyping:** following our recent success in the prototyping of GNNs for computer network applications [60], we would integrate the acquired knowledge into openly shared implementations, conceived with the plan of future integration into an open GRAPHSEC framework, providing GNN prototyping capabilities tailored to AI4SEC applications.

④ **Proof of Concept:** we would demonstrate GRAPHSEC technology in different AI4SEC applications, of different nature in terms of supervision guidance (supervised, self-supervised, and unsupervised), data-representation size (scalability trade-offs), and data dynamics.

We expect the following results from the GRAPHSEC research agenda, driven-by and applicable-to different AI4SEC applications: (a) graph design, construction, and embedding methodologies specific to network security data (e.g., net2graph, ip2graph, flow2graph, proto2graph); (b) GNN-based methods for graph modeling and learning, oriented to (static and dynamic) AI4SEC; (c) systematic comparison of GNN-based AI4SEC with respect to state-of-the-art AI/ML approaches, in terms of task performance, generalization, scalability, and robustness; (d) GRAPHSEC proof-of-concept (PoC) implementations for targeted applications; (e) graph-structured research datasets available to the scientific community, serving research reproducibility and benchmarking purposes.

4 Conclusions

The ultimate goal of this positioning paper is to contribute to strengthening the future research on the AI4SEC field, by pinpointing some of the fundamental challenges it faces, proposing novel technology to improve the state of affairs. We believe that a fundamental limitation is the lack of AI/ML technology specific to network security. The properties and initial results obtained in the application of GNNs to multiple domains make us believe that graph-structured representations and GNNs could be foundational to AI4SEC, in the way convolutional and recursive networks were to CVSR problems. We acknowledge that the discussion presented in this paper is by no means fully exhaustive and final, and we are sure many aspects were left aside; however, we do hope this paper will motivate further discussion on the limitations we are facing in AI4SEC, and help in making of GNNs and graph ML powerful enablers for better AI4SEC.

Acknowledgements. This work is partially funded by the Austrian FFG ICT-of-the-Future DynAISEC project (ref. 887504 – *Adaptive AI/ML for Dynamic Cybersecurity Systems*).

References

1. LeCun, Y., Bengio, Y., Hinton, G.E.: Deep learning. Nature **521**(7553), 436–444 (2015)
2. Bronstein, M.M., Bruna, J., LeCun, Y., Szlam, A., Vandergheynst, P.: Geometric deep learning: going beyond euclidean data. IEEE Sig. Process. Mag. **34**(4), 18–42 (2017)
3. Scarselli, F., et al.: The graph neural network model. IEEE Trans. NNets. **20**(1), 61–80 (2009)
4. Battaglia, P.W., et al.: Relational Inductive Biases, Deep Learning, and Graph Networks. arXiv preprint arXiv:1806.01261 (2018)
5. Zhou, J., et al.: Graph neural networks: a review of methods and applications. AI Open **1**, 57–81 (2020)
6. Garg, V.K., Jegelka, S., Jaakkola, T.S.: Generalization and representational limits of graph neural networks. In: 37th International Conference on Machine Learning (ICML). vol. 119, pp. 3419–3430. PMLR (2020)
7. Zhu, Q., Ponomareva, N., Han, J., Perozzi, B.: Shift-Robust GNNs: overcoming the limitations of localized graph training data. In: 34th Advances in Neural Information Processing Systems (NeurIPS), pp. 27965–27977 (2021)
8. Xu, K., Hu, W., Leskovec, J., Jegelka, S.: How powerful are graph neural networks?. In: 7th International Conference on Learning Representations (ICLR), OpenReview.net (2019)
9. Halcrow, J., Mosoi, A., Ruth, S., Perozzi, B.: Grale: designing networks for graph learning. In: KDD 2020: Proceedings of the 26th ACM SIGKDD International Conference on Knowledge Discovery & Data Mining, pp. 2523–2532. ACM (2020)
10. Serafini, M.: Scalable graph neural network training: the case for sampling. ACM SIGOPS Oper. Syst. Rev. **55**(1), 68–76 (2021)
11. Suarez-Varela, J., et al.: Graph neural networks for communication networks: context, use cases and opportunities. IEEE Netw. 1–8 (2022)
12. Barcelona Neural Networking Center. Must read papers on GNN for communication networks. https://github.com/BNN-UPC/GNNPapersCommNets. Accessed 10 Oct 2022
13. Rusek, K., et al.: RouteNet: leveraging graph neural networks for network modeling and optimization in SDN. IEEE J. Sel. Areas Commun. **38**(10), 2260–2270 (2020)
14. Sommer, R., Paxson, V.: Outside the closed world: on using machine learning for network intrusion detection. In: 31st 2010 IEEE Symposium on Security and Privacy, pp. 305–316. IEEE (2010)
15. Carela-Español, V., Barlet-Ros, P., Bifet, A., Fukuda, K.: A streaming flow-based technique for traffic classification applied to 12 + 1 years of Internet traffic. Telecommun. Syst. **63**(2), 191–204 (2015). https://doi.org/10.1007/s11235-015-0114-6
16. Domingos, P., Hulten, G.: Mining high-speed data streams. In: ACM SIGKDD International Conference on Knowledge Discovery and Data Mining (2000)

17. Casas, P., Mulinka, P., Vanerio, J.: Should I (re)Learn or Should I Go(on)? Stream machine learning for adaptive defense against network attacks. In: 26th ACM Conference on Computer and Communications Security (CCS), 6th ACM Workshop on Moving Target Defense, MTD (2019)

18. Boutaba, R., et al.: A comprehensive survey on machine learning for networking: evolution, applications and research opportunities. J. Internet Serv. Appl. **9**(1), 16:1–16:99 (2018)

19. Chandola, V., Banerjee, A., Kumar, V.: Anomaly detection: a survey. ACM Comput. Surv. **41**(3), 15:1–15:58 (2009)

20. Ahmed, M., Mahmood, A.N., Hu, J.: A survey of network anomaly detection techniques. J. Netw. Comput. Appl. **60**, 19–31 (2016)

21. Mahdavifar, S., Ghorbani, A.A.: Application of deep learning to cybersecurity: a survey. Neurocomputing **347**, 149–176 (2019)

22. Saxe, J., Harang, R.E., Wild, C., Sanders, H.: A Deep Learning Approach to Fast, Format-Agnostic Detection of Malicious Web Content. arXiv preprint arXiv:1804.05020 (2018)

23. Radford, B.J., Apolonio, L.M., Trias, A.J., Simpson, J.A.: Network Traffic Anomaly Detection Using Recurrent Neural Networks. arXiv preprint arXiv:1803.10769 (2018)

24. Zhao, J., Shetty, S., Pan, J.W.: Feature-based transfer learning for network security. In: 2017 IEEE Military Communications Conference (MILCOM), pp. 17–22. IEEE (2017)

25. Guo, W., et al.: LEMNA: Explaining deep learning based security applications. In: Proceedings of the 2018 ACM SIGSAC Conference on Computer and Communications Security, pp. 364–379. ACM (2018)

26. Zeng, Z., et al.: Building more explainable artificial intelligence with argumentation. In: Proceedings of the 32nd Conference on Artificial Intelligence (AAAI) (2018)

27. Miller, T.: Explanation in artificial intelligence: insights from the social sciences. Artif. Intell. **267**, 1–38 (2019)

28. Gunning, D.: Explainable Artificial Intelligence (XAI). Defense Advanced Research Projects Agency (DARPA) (2017). https://www.darpa.mil/attachments/XAIProgramUpdate.pdf

29. Hazard, C.J., et al.: Natively Interpretable Machine Learning and Artificial Intelligence: Preliminary Results and Future Directions. arXiv preprint arXiv:1901.00246 (2019)

30. Ribeiro, M.T., Singh, S., Guestrin, C.: Why should i trust you?: Explaining the predictions of any classifier. In: ACM SIGKDD International Conference on Knowledge Discovery and Data Mining (2016)

31. Lundberg, S.M., Lee, S.-I.: A unified approach to interpreting model predictions. In: Proceedings of the 31st International Conference on Neural Information Processing Systems (NIPS) (2017)

32. Kuppa, A., Le-Khac, N.-A.: Black box attacks on explainable artificial intelligence (XAI) methods in cyber security. In: International Joint Conference on Neural Networks (IJCNN) (2020)

33. Shejwalkar, V., et al.: Back to the drawing board: a critical evaluation of poisoning attacks on production federated learning. In: 43rd 2022 IEEE Symposium on Security and Privacy (SP), pp. 1354–1371. IEEE (2022)

34. Ring, M., et al.: A Survey of Network-based Intrusion Detection Data Sets. arXiv preprint arXiv:1903.02460 (2019)

35. Trevisan, M., et al.: Five years at the edge: watching internet from the ISP network. In: IEEE/ACM Transactions on Networking. vol. 28(2) (2020)
36. Labovitz, C.: Internet Traffic 2009–2019. Presentation at NANOG 76 (2019)
37. Glass-Vanderlan, T., et al.: A survey of intrusion detection systems leveraging host data. In: ACM Computing Surveys. vol. 52(6) (2019)
38. Molnár, S., Megyesi, P., Szabó, G.: How to validate traffic generators. In: Proceedings of the IEEE International Conference on Communications Workshops (ICC-W) (2013)
39. Erlacher, F., Dressler, F.: How to test an IDS? GENESIDS: an automated system for generating attack traffic. In: Proceedings of the Workshop on Traffic Measurements for Cybersecurity (WTMC) (2018)
40. Ring, M., et al.: Flow-basd Network Traffic Generation using Generative Adversarial Networks. arXiv preprint arXiv:1810.07795 (2018)
41. Park, N., et al.: Data synthesis based on generative adversarial networks. In: Proceedings of the VLDB Endowment (2018)
42. Goodfellow, I., et al.: Generative adversarial nets. In: Advances in Neural Information Processing Systems. vol. 27 (2014)
43. Brock, A., Donahue, J., Simonyan, K.: Large Scale GAN Training for High Fidelity Natural Image Synthesis. arXiv preprint arXiv:1809.11096 (2018)
44. Shorten, C., Khoshgoftaar, T.M.: A survey on image data augmentation for deep learning. J. Big Data 6(1), 1–48 (2019). https://doi.org/10.1186/s40537-019-0197-0
45. Zhou, D., Bousquet, O., Lal, T.N., Weston, J., Scholkopf, B.: Learning with local and global consistency. In: Advances in Neural Information Processing Systems. vol. 16 (2004)
46. Zhu, X., Ghahramani, Z.: Learning from labeled and unlabeled data with label propagation. Carnegie Mellon University, Tech. Rep. (2002). CMU-CALD-02-107
47. Li, G., et al.: Deep learning algorithms for cyber security applications: a survey. J. Comput. Secur. 29(5), 447–471 (2021)
48. Cui, P., et al.: A survey on network embedding. IEEE Trans. Knowl. Data Eng. 31, 833–852 (2019)
49. Hamilton, W.L., et al.: Inductive representation learning on large graphs. In: 30th Advances in Neural Information Processing Systems (NIPS) (2017)
50. Barceló, P., et al.: The logical expressiveness of graph neural networks. In: 8th International Conference on Learning Representations, ICLR 2020, Addis Ababa, Ethiopia, 26–30 April 2020. OpenReview.net (2020)
51. Zhang, X., Zitnik, M.: GNNGuard: defending graph neural networks against adversarial attacks. In: 33rd NeurIPS (2020)
52. Xie, Y., Xu, Z., Wang, Z., Ji, S.: Self-Supervised Learning of Graph Neural Networks: A Unified Review. arXiv preprint arXiv:2102.10757 (2021)
53. Trivedi, R., et al.: Representation Learning over Dynamic Graphs. arXiv preprint arXiv:1803.04051 (2018)
54. Rossi, E., et al.: Temporal Graph Networks for Deep Learning on Dynamic Graphs. arXiv preprint arXiv:2006.10637 (2020)
55. Palowitch, J., Tsitsulin, A., Mayer, B., Perozzi, B.: GraphWorld: fake graphs bring real insights for GNNs. In: KDD 2022: Proceedings of the 28th ACM SIGKDD Conference on Knowledge Discovery and Data Mining, pp. 3691–3701. ACM (2022)
56. Xu, P., Eckert, C., Zarras, A.: Detecting and categorizing android malware with graph neural networks. In: SAC 2021: Proceedings of the 36th Annual ACM Symposium on Applied Computing, pp. 409–412. ACM (2021)

57. Busch, J., et al.: NF-GNN: network flow graph neural networks for malware detection and classification. In: SSDBM 2021: 33rd International Conference on Scientific and Statistical Database Management, pp. 121–132. ACM (2021)
58. Zhou, J., Xu, Z., Rush, A.M., Yu, M.: Automating Botnet Detection with Graph Neural Networks. arXiv preprint arXiv:2003.06344 (2020)
59. Pujol-Perich, D., Suárez-Varela, J., Cabellos-Aparicio, A., Barlet-Ros, P.: Unveiling the Potential of Graph Neural Networks for Robust Intrusion Detection. arXiv preprint arXiv:2107.14756 (2021)
60. Pujol-Perich, D., et al.: IGNNITION: bridging the gap between graph neural networks and networking systems. IEEE Netw. **35**(6), 171–177 (2021)

Low Complexity Adaptive ML Approaches
for End-to-End Latency Prediction

Pierre Larrenie[1,2]([mail]) [iD], Jean-François Bercher[2], Iyad Lahsen-Cherif[3] [iD],
and Olivier Venard[4] [iD]

[1] Thales SIX, Gennevilliers, France
pierre.larrenie@thalesgroup.com
[2] LIGM, Univ. Gustave Eiffel, CNRS, Marne-la-Vallée, France
{pierre.larrenie,jean-francois.bercher}@esiee.fr
[3] Institut National des Postes et Télécommunications (INPT), Rabat, Morocco
lahsencherif@inpt.ac.ma
[4] ESYCOM, Univ. Gustave Eiffel, CNRS, Marne-la-Vallée, France
olivier.venard@esiee.fr

Abstract. Software Defined Networks have opened the door to statistical and AI-based techniques to improve efficiency of networking. Especially to ensure a certain *Quality of Service* (QoS) for specific applications by routing packets with awareness on content nature (VoIP, video, files, etc.) and its needs (latency, bandwidth, etc.) to use efficiently resources of a network.

Monitoring and predicting various Key Performance Indicators (KPIs) at any level may handle such problems while preserving network bandwidth.

The question addressed in this work is the design of efficient, low-cost adaptive algorithms for KPI estimation, monitoring and prediction. We focus on end-to-end latency prediction, for which we illustrate our approaches and results on data obtained from a public generator provided after the recent international challenge on GNN [12].

In this paper, we improve our previously proposed low-cost estimators [6] by adding the adaptive dimension, and show that the performances are minimally modified while gaining the ability to track varying networks.

Keywords: KPI Prediction · Machine Learning · Adaptivity · General Regression · SDN · Networking

1 Introduction

Routing while ensuring quality of service (QoS) remains a significant challenge in all networks. Whatever the resources, their use must be optimized to satisfy both throughput and QoS to users. This is true for static wide area networks, but even more so for mobile networks with dynamic topology.

The emergence of software-defined networks (SDNs) [1,11] has enabled data to be shared more efficiently across communication layers. Services can provide

É. Renault and P. Mühlethaler (Eds.): MLN 2022, LNCS 13767, pp. 72–87, 2023.
https://doi.org/10.1007/978-3-031-36183-8_6

network requirements to routers; routers acquire data about network performance and allocate resources to meet those requirements as best as possible. However, acquiring overall network performance can result in high network bandwidth consumption for signaling, degrading the available resources, and is particularly limiting for resource-constrained networks such as mobile networks (MANETs).

We consider a network for which we wish to reduce signaling and perform intelligent routing. In order to limit the amount of signaling, the first axis is to estimate some key performance indicators (KPIs) from other KPIs. A second axis would be to perform this prediction locally, at the node level, rather than a global estimation in the network. Finally, if predictions are to be performed locally, the complexity of the algorithms will need to be low while preserving good prediction quality. The last point is to be able to detect and track changes in the state of the network, which implies that the predictors will have to use only a small number of the previous states of the network and be able to readapt continuously.

The question addressed in this work is the design of efficient, low-cost adaptive algorithms for KPI estimation, monitoring and prediction. In the present paper, we improve our previously proposed low-cost estimators [6] by adding the adaptive dimension and show that the performances are minimally modified while gaining the ability to track varying networks. We focus on end-to-end latency prediction, for which we illustrate our approaches and results on data we generated using a public generator made available after the recent international challenge [12]. The best performances of the state-of-the-art are obtained with Graph Neural Networks (GNNs) [3,10,12]. Although this is a global method while we favor local and adaptive methods, we used these performances as a benchmark.

We present related works in Sect. 2. Then we present in Sect. 3 our main results from [6]. Instead of using high performances global and high-costs methods based on Graph Neural Networks (GNNs) [3,10,12], we proposed to use standard machine learning regression methods. We showed that a careful feature engineering and feature selection (based on queue theory and the approach in [2]), as well as the use of a single feature with curve-fitting methods, allows to obtain near state-of-the-art performances with both a very low number of parameters, significantly lower learning and inference times compared to GNNs, and the with the ability to operate at the link level instead of a whole-graph level. In Sect. 4, we show how these block algorithms can be transformed into versions implementable in an iterative way (i.e. by taking into account the data one by one as they become available), with the originality of using a regularization term. Then, time dependent estimations, or the addition of forgetting factor will give them an adaptive character. In Sect. 5 we describe the validation dataset we built from a public generator and then the results of our experimentation. Finally, we conclude, discuss the overall results and draw some perspectives.

2   Related Work

[4] present an heuristic and an Mixed Integer Programming approach to optimize Service Functions Chain provisioning when using Network Functions

Virtualization for a service provider. Their approach relies on minimizing a trade-off between the expected latency and infrastructures resources.

Such optimization routing flow in SDN may need additional information to be exchanged between the nodes of a network. This results in an increase of the volume of signalization, by performing some measurements such as in [7]. This is not a consequent problem in unconstrained networks, i.e. static wired networks with near-infinite bandwidth but may decrease performance of wireless network with poor capacity. An interesting solution to save bandwidth would be to predict some of the KPIs from other KPIs and data exchanged globally between nodes.

In [8,9], authors proposed a MANETs application of SDN in the domain of tactical networks. They proposed a multi-level SDN controllers architecture to build both secure and resilient networking. While orchestrating communication efficiently under military constraints such as: high-level of dynamism, frequent network failures, resources-limited devices. The proposed architecture is a trade-off between traditional centralized architecture of SDN and a decentralized architecture to meet dynamic in-network constraints.

[5] proposed a Quality of Experience (QoE) management strategy in a SDN to optimize the loading time of all the tile of a mapping application. They have shown the impact of several KPIs on their application using a Generalized Linear Model (GLM). This mechanism make the application aware of the current network state.

[10] used GNNs for predicting KPIs such as latency, error-rate and jitter. They relied on the *Routenet* architecture of Fig. 1. The idea is to model the problem as a bipartite hypergraph mapping flows to links as depicted on Fig. 2. Aggregating messages in such graph may result in predicting KPIs of the network in input. The model needs to know the routing scheme, traffic and links properties. Their result is very promising and has been the subject of two ITU Challenge in 2020 and 2021 [3,12]. These ITU challenges have very good results since the top-3 teams are around 2% error in delay prediction in the sense of Mean-Absolute Percentage Error (MAPE).

In [2], very promising results were obtained with a a near 1% GNN model error (in the sense of MAPE) on the test set. The model mix analytical $M/M/1/K$ queueing theory used to create extra-features to feed GNN model. In order to satisfy the constraint of scalability proposed by the challenge, the first part of model operates at the link level.

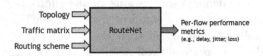

Fig. 1. Routenet Architecture [10]

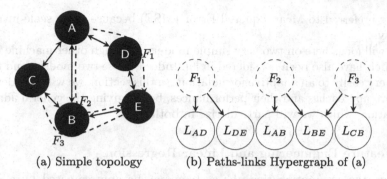

(a) Simple topology   (b) Paths-links Hypergraph of (a)

Fig. 2. Routenet [10] paths-links hypergraph transformation applied on a simple topology graph carrying 3 flows. (a) Black circles represents communication node, double headed arrows between them denotes available symmetric communications links and dotted arrows shows flows path. (b) Circle (resp. dotted) represents links (resp. flows) entities defined in the first graph (L_{ij} is the symmetric link between node i and node j.). Unidirectional arrows encode the relation "<flow> is carried by <link>".

3   Simple Machine-Learning Approaches for Latency Prediction

Our first problem is to define an estimator $\hat{y}$ of the occupancy y as a function of the different available "features" of the system, with a joint objective of low complexity and performance. To do so, we will look for an approximation function $f_\theta(\mathbf{u})$ that allows to estimate y from the features $\mathbf{u}$ and parameters $\boldsymbol{\theta}$.

$$\hat{y} = f_\theta(\mathbf{u}) \tag{1}$$

Here $\mathbf{u}$ and $\boldsymbol{\theta}$ are vectors that collect the different features or parameters. Once an estimate of occupancy is obtained, it is possible to get the latency prediction $\hat{d}_n$ for a specific link n by the simple relation

$$\hat{d}_n = \hat{y}_n \frac{\mathbb{E}(|P_n|)}{c_n} \tag{2}$$

where $\mathbb{E}[|P_n|]$ is the observed average packet size on link n and c_n the capacity of this link.

For analytical simplicity, the parameters $\boldsymbol{\theta}$ will be sought by minimizing the minimum mean square error

$$\mathbb{E}\left[(y - \hat{y})^2\right] = \mathbb{E}\left[(y - f_\theta(\mathbf{u}))^2\right], \tag{3}$$

although the performances are also often evaluated in the MAPE sense

$$\mathcal{L}(\hat{y}, y) = \frac{100\%}{N} \sum_{n=1}^{N} \left| \frac{\hat{y}_n - y_n}{y_n} \right| \tag{4}$$

which is preferred to Mean Squared Error (MSE) because of its scale-invariant property.

We will focus here on two very simple models, although other machine learning models have also been considered in [6]. Indeed, these two models lend themselves very easily to an adaptive formulation. In this section, we will first describe these two approaches and their performances, before giving the general adaptive formulation, which we will particularize in both cases.

3.1   Feature Engineering and Linear Regression

Based on the assumption that the system may be approximated by a model whose essential features come from $M/M/1/K$ and $M/G/1/K$ queue theory, we took essential parameters characterizing queueing systems, such as: ρ, ρ_e, π_0, π_K, etc. and built further features by applying interactions and various non-linearities (powers, log, exponential, square root). Then, we selected features in this set by a forward step-wise selection method; i.e. by adding in turn each feature to potential models and keeping the feature with best performance. Finally, we selected the model with best MAPE error. For a linear regression model, this led us to select and keep a set of 4 simple features, which interestingly enough, have simple interpretations:

$$\begin{cases} \pi_0 = \frac{1-\rho}{1-\rho^{K+1}} \\ L = \rho + \pi_0 \sum_k k\rho^k \\ \rho_e = \frac{\lambda_e}{\lambda}\rho = \frac{\lambda_e}{\mu} \\ S_e = \sum_k k\rho_e^k \end{cases} \tag{5}$$

where L is the expected number of packets in the queue according to $M/M/1/K$, π_0 the probability that the queue is empty according to $M/M/1/K$ theory, ρ_e the effective queue utilization, and S_e the unnormalized expected value of the effective number of packet in the queue buffer. These features can be thought as a kind of data preprocessing, before applying ML algorithms, and this turns out to be a key to achieving good performances. The 4 previous features have been used as input for several machine learning models like Multi-Layer Perceptron model (MLP), Linear Regression, SVM, Random Forest, Gradient Boosting Regression Tree. We only describe here the case of linear regression, since it is a method for which an adaptive version is readily obtained. In this case, model (1) is simply

$$\hat{y} = \theta_0 + \theta_1\pi_0 + \theta_2 L + \theta_3\rho_e + \theta_4 S_e = \boldsymbol{\theta}^T \mathbf{u} \tag{6}$$

with $\boldsymbol{\theta}^T = [\theta_0, \ldots \theta_4]$ and $\mathbf{u}^T = [1, \pi_0, L, \rho_e, S_e]$. For the linear regression model in ((6), it is well known that the regularized minimum mean squared error

$$J(\boldsymbol{\theta}) = \mathbb{E}\left[(y - \boldsymbol{\theta}^T \mathbf{u})^2\right] + \alpha||\boldsymbol{\theta}||^2 \tag{7}$$

is obtained for

$$\boldsymbol{\theta} : (\mathbf{R}_{uu} + \alpha\mathbf{1})\,\boldsymbol{\theta} = \mathbf{R}_{yu} \tag{8}$$

where we denoted

$$\begin{cases} \mathbf{R}_{uu} = \mathbb{E}\left[\mathbf{u}\mathbf{u}^T\right], & \text{the correlation matrix of } \mathbf{u} \\ \mathbf{R}_{yu} = \mathbb{E}\left[y\mathbf{u}\right], & \text{the correlation vector of } y \text{ and } \mathbf{u} \end{cases}$$

and $\mathbf{1}$ the identity matrix, α the regularization parameter.

As far as performance is concerned with this approach, it was evaluated using static data from the GNN ITU Challenge 2021 [12]. Compared to the state-of-the-art, our linear regression with carefully selected features shows a very slight performance degradation: 1.74% in MAPE while the best state-of-the-art method is at 1.27%. One strong advantage is in term of training and inference time. It has a training time of less than a second when GNN requires more than 8 h. Moreover, the inference time for the complete network is also much lower, by a factor of almost 1000 (0.296s vs 214s).

3.2 Curve Regression by Bernstein Polynomials

There is a high interdependence of the features we selected in Equation (5), since all these features can be expressed in term of ρ_e. Furthermore, it is confirmed by data exploration that ρ_e is the prominent feature for occupancy prediction (and in turn latency prediction), as exemplified in Fig. 3.

It is then tempting to try to further simplify our features space and estimate the occupancy from a non-linear transformation of the single feature ρ_e, as:

$$\hat{y} = g(\rho_e) \tag{9}$$

where $\hat{y}$ is the estimate of the occupancy y. The concerns are of course to define simple and efficient functions g, with a low number of parameters, that can model the kind of growth shown in Fig. 3, and of course to check that the performance remains interesting.

The estimator g is defined as a linear combination of simple functions f_n:

$$\hat{y} = g(\rho_e) = \sum_n \theta_n \cdot f_n(\rho_e) \tag{10}$$

which is also a linear model in terms of function $f_n(\rho_e)$.

Several solutions were considered in [6] to define or choose the functions f_n. Since we know that the Bernstein polynomials form a basis in the set of polynomial in the interval $[0;1]$; and that the approximation of any continuous function on $[0;1[$ by a Bernstein polynomial converges uniformly, we were led to these polynomials:

$$f_n^K(x) = \binom{K}{n} x^n (1-x)^{K-n} \tag{11}$$

where K is maximum order of polynomials.

As mentioned, (10) can be rewritten as the linear model

$$\hat{y} = g(\rho_e) = \sum_n \theta_n \cdot f_n(\rho_e) = \boldsymbol{\theta}^T \mathbf{u} \tag{12}$$

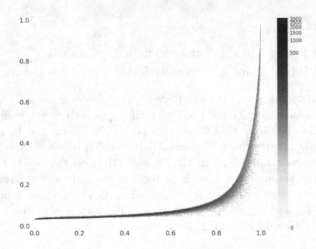

Fig. 3. Data of ITU Challenge 2021 [12], ρ_e vs queue occupancy. Color-scale is an indicator of points cloud density.

with $\boldsymbol{\theta}^T = [\theta_0, \dots, \theta_K]$ and $\mathbf{u}^T = [f_0^K(\rho_e), f_1^K(\rho_e), \dots, f_K^K(\rho_e)]$. Hence, we have the same form as in (8) for the solution.

In term of performances, we also obtained a minor degradation in MAPE (1.68%) compared to state-of-the-art (1.29%), while improving by several orders the wall training and inference times (2min/3.14s vs 8hrs/214s); though a bit less than the simple linear regression.

4   Adaptive Versions

We place ourselves in the context where we have regular snapshots of the state of the network, which allows us to both monitor the quality of predictions, and to track changes in the network. For the n-th series of measurements, let us denote $y(n)$ the measured latency and $\mathbf{u}(n)$ the features. We can also group several snapshots or several links into a vector of latencies $\mathbf{y}(n)$ and matrix $\mathbf{U}(n)$. In the following we will derive equations for this block case, which includes immediately the scalar case.

The minimum mean square error (7) which has the explicit solution (8) can also be solved by a gradient algorithm as

$$\boldsymbol{\theta}_{k+1} = \boldsymbol{\theta}_k - \mu \, \nabla J(\boldsymbol{\theta})|_{\boldsymbol{\theta}=\boldsymbol{\theta}_k}, \tag{13}$$
$$= \boldsymbol{\theta}_k - \mu \left((\mathbf{R}_{uu} + \alpha \mathbf{1}) \, \boldsymbol{\theta}_k - \mathbf{R}_{yu} \right). \tag{14}$$

In (14), we can substitute the true values with estimated ones. In order to introduce adaptivity to context changes in the network, these estimates will preserve the temporal dimension. We thus use either a sliding average

$$\begin{cases} \hat{\mathbf{R}}_{uu}(n) = \sum_{l=0}^{L} \mathbf{U}(n-l)\mathbf{U}(n-l)^T \\ \hat{\mathbf{R}}_{yu}(n) = \sum_{l=0}^{L} \mathbf{U}(n-l)\mathbf{y}(n-l) \end{cases} \tag{15}$$

or an exponential mean

$$\begin{cases} \hat{\mathbf{R}}_{uu}(n) = \sum_{l=0}^{n} \lambda^{l-n} \mathbf{U}(l)\mathbf{U}(l)^T = \lambda\hat{\mathbf{R}}_{uu}(n-1) + \mathbf{U}(n)\mathbf{U}(n)^T \\ \hat{\mathbf{R}}_{yu}(n) = \lambda\hat{\mathbf{R}}_{yu}(n-1) + \mathbf{U}(n)\mathbf{y}(n). \end{cases} \tag{16}$$

where $\lambda \leq 1$ is the forgetting factor.

In the limit case where we take either $L = 0$ or $\lambda = 0$ in the previous formulas, we get the 'instantaneous estimates'

$$\begin{cases} \hat{\mathbf{R}}_{uu}(n) = \mathbf{U}(n)\mathbf{U}(n)^T \\ \hat{\mathbf{R}}_{du}(n) = \mathbf{U}(n)\mathbf{y}(n). \end{cases} \tag{17}$$

we obtain

$$\boldsymbol{\theta}(n+1) = (1 - \mu\alpha)\boldsymbol{\theta}(n) - \mu\mathbf{U}(n)\left(\mathbf{U}(n)^T\boldsymbol{\theta}(n) - \mathbf{y}(n)\right) \tag{18}$$

which reduces to the well known LMS algorithm [14] in the scalar case and no regularization, $\alpha = 0$.

Alternatively, one can try to solve the normal equation (8), using the time dependent estimates as the exponential mean (16). The difficulty with the solution

$$\hat{\boldsymbol{\theta}}(n+1) = \left[\hat{\mathbf{R}}_{uu}(n+1) + \alpha\mathbf{1}\right]^{-1} \hat{\mathbf{R}}_{yu}(n+1) \tag{19}$$

is the inversion, for each n, of the correlation matrix. Let us denote

$$\mathbf{K}(n+1) = \left[\hat{\mathbf{R}}_{uu}(n+1) + \alpha\mathbf{1}\right]^{-1}. \tag{20}$$

Using (16), we have

$$\mathbf{K}(n+1)^{-1} = \lambda\hat{\mathbf{R}}_{uu}(n) + \mathbf{U}(n+1)\mathbf{U}(n+1)^T + \alpha\mathbf{1} \tag{21}$$

$$= \lambda\left(\hat{\mathbf{R}}_{uu}(n) + \alpha\mathbf{1}\right) + \mathbf{U}(n+1)\mathbf{U}(n+1)^T + \alpha(1 - \lambda)\mathbf{1} \tag{22}$$

$$= \lambda\mathbf{K}(n)^{-1} + \mathbf{U}(n+1)\mathbf{U}(n+1)^T + \alpha(1 - \lambda)\mathbf{1} \tag{23}$$

and

$$\mathbf{K}(n+1) = \left[\left(\lambda\mathbf{K}(n)^{-1} + \mathbf{U}(n+1)\mathbf{U}(n+1)^T\right) + \alpha(1-\lambda)\mathbf{1}\right]^{-1} \tag{24}$$

$$= \left[\mathbf{Q}(n+1) + \delta\mathbf{1}\right]^{-1} \tag{25}$$

with

$$\mathbf{Q}(n+1) = \left(\lambda\mathbf{K}(n)^{-1} + \mathbf{U}(n+1)\mathbf{U}(n+1)^T\right) \tag{26}$$

and $\delta = \alpha(1 - \lambda)$ The matrix inversion lemma enables to reduce the inversion of $\mathbf{Q}(n+1)$ to

$$\mathbf{Q}(n+1)^{-1} = \frac{1}{\lambda}\mathbf{K}(n) - \frac{1}{\lambda^2}\mathbf{K}(n)\mathbf{U}(n+1)\times$$
$$\left(1 + \frac{1}{\lambda}\mathbf{U}(n+1)^T\mathbf{K}(n)\mathbf{U}(n+1)\right)^{-1}\mathbf{U}(n+1)^T\mathbf{K}(n), \tag{27}$$

which simplifies to

$$\mathbf{Q}(n+1)^{-1} = \frac{1}{\lambda}\mathbf{K}(n) - \frac{1}{\lambda^2}\frac{\mathbf{K}(n)\mathbf{u}(n+1)\mathbf{u}(n+1)^T\mathbf{K}(n)}{1 + \frac{1}{\lambda}\mathbf{u}(k+1)^T\mathbf{K}(n)\mathbf{u}(k+1)}, \tag{28}$$

for scalar observations. Now, we can use the Taylor expansion to get

$$\mathbf{K}(n+1) = [\mathbf{Q}(n+1) + \delta\mathbf{1}]^{-1} = \mathbf{Q}(n+1)^{-1} - \delta\mathbf{Q}(n+1)^{-2} + \delta^2\mathbf{Q}(n+1)^{-3} + \cdots \tag{29}$$

This gives us a way to compute recursively the inverse of the regularized estimate of the correlation matrix by combining (27) and (29) into

$$\mathbf{K}(n+1) \approx \mathbf{Q}(n+1)^{-1} - \delta\mathbf{Q}(n+1)^{-2} \tag{30}$$

which, by (27), does not require the inversion of $K(n)$.

In both cases, we have the updating formula

$$\boldsymbol{\theta}(n+1) = \boldsymbol{\theta}(n) + \mathbf{K}(n+1)\mathbf{U}(n+1)[\mathbf{y}(n+1) - \boldsymbol{\theta}(n)^T\mathbf{U}(n+1)]. \tag{31}$$

5 Experiments and Results

5.1 Dataset

We generate a dataset thanks to a public challenge data generator [12]. This data generator is based on the well-known OMNET++ discrete event network simulator [13]. The published simulator is available as a docker image. However, due to the rules of the 2022 edition of the challenge, it is not possible to generate large topologies, i.e. no more than 10 nodes. Since our models are link-based, the use of small topologies does not seem to be a problem. The simulator is parameterized by a traffic matrix and a topological graph that are easy to generate thanks to the provided API.

Our generated dataset, used is this paper, is the result of 11900 simulations of the same topology graph of 10 nodes and 30 links, subject to 100 different traffic matrices. In order to get complex results of simulations but at low cost, we made the choice to model a network with small queue buffers (8000 bits) and possibly subject of high traffic intensities (maximum traffic rate set to 4000 bits/s for each flow). Then for each traffic matrices, we alter the capacity of the network according to a sigmoid, in order to model a network subject to jamming, with 2 stationary states. The proposed jamming may cause a decrease in the capacity of the network links by up to a factor of 5, as depicted on Fig. 4a. For simplification purposes, we have considered that each link of the network has the same capacity. This result in a U-shaped distribution of our link data samples according to the link capacity as shown in Fig. 4b.

5.2 Results

From the generated data, we validate our approach along several axes.

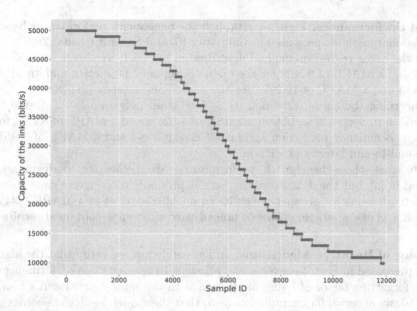

(a) Capacity alteration to model jamming with a decrease of the capacity up to a factor of 5.

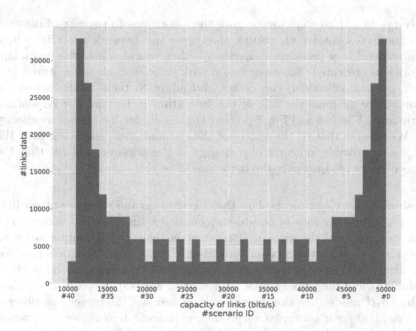

(b) Link capacity distribution of the generated dataset.

Fig. 4. Overview of the generated dataset

Global Performances. First, we establish the benchmark performances based on the global methods presented in the paper [6] and Sects. 3.1 and 3.2.

For the linear regression method described in Sect. 3.1, we obtain an MSE of 5.86e-4 and a MAPE of 9.58% for the queue occupancy prediction and an MSE of 1.10e-3 and a MAPE of 10.26%. for the end-to-end latency prediction.

Concerning the curve regression using Bernstein polynomials (of degree 8) described in in Sect. 3.2, we obtain an MSE of 4.52e-4 and a MAPE of 8.72% for the queue occupancy prediction and an MSE of 9.35e-4 and a MAPE of 9.95% for the end-to-end latency prediction.

Note that these benchmark performances are below the performances obtained in [6], but the dataset we use here is probably more severe since using ground truth value for occupancy results in an MSE 6.03e-4 of a MAPE 9.34% for the flow delay prediction. That is indeed very close of the obtained results.

Behavior of Iterative Algorithms. In a second step, we verify that the algorithms presented in Sect. 4 converge and allow us to recover these performances. With a forgetting factor of 1 (use of all data with the same weight) and a block size of 10, we observe, for example in Fig. 5, that the model coefficients converge towards a stable value, and that MAPE recovers the value obtained with the global method using all data. The convergence is obtained in less than 10,000 operations. It is thus possible to replace the global method, which is already low-cost, with an approach where the calculations are carried out recursively.

Adaptivity. In a third step, we compare the algorithms to the case of network modifications We consider an abrupt change in the network capacity, which could correspond to a jamming scenario. We then examine how the two adaptive algorithms presented (linear regression with judiciously chosen features; and Bernstein polynomial model) can detect and adapt to these modifications. In this context, we examine the role of the forgetting factor and the regularization parameter. Figure 6 and Fig. 7 present the results for the case of a capacity change. We observe that (i) the square of the residual error, smoothed over 100 points, is a remarkable indicator of a change in the network; and (ii) that the model coefficients readjust over the iterations after this change.

Discussion. These experiments show the effectiveness and relevance of our iterative and adaptive versions of end-to-end latency estimation procedures. The iterative versions have the same performance as their global counterparts; an even lower cost since they can be implemented iteratively as the data is received or made available. The convergence time for the model coefficients is a few thousand samples while the global model used around 350,000 samples for training, while the GNN models require several million samples. Moreover, we observe that the residual error converges very quickly, in some tens of samples, which means that although the convergence of the models' coefficients does not seem to be complete, they are equivalent from the point of view of performance for occupancy prediction. From an operational point of view, the model can be refreshed

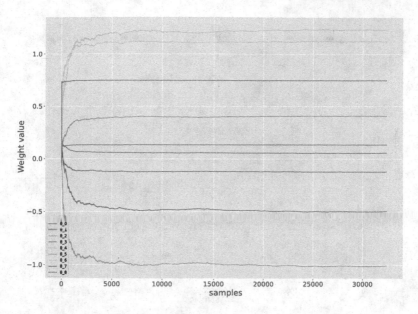

(a) Iterative curve-fitting based on Bernstein polynomials of degree 8.

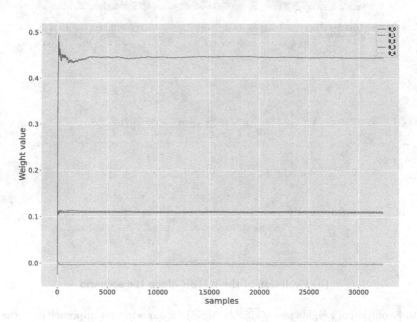

(b) Iterative Linear Regression

Fig. 5. Evolution of weights for our iterative methods without forgetting (non-adaptive) while fitting the whole dataset.

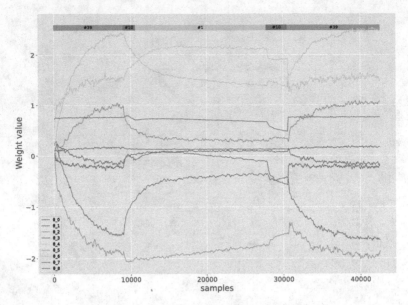

(a) Evolution of weights along the scenario.

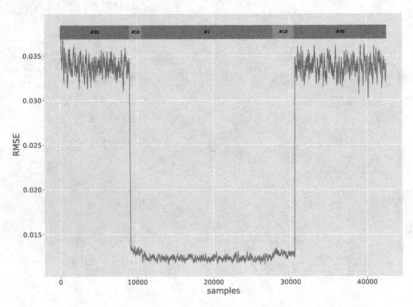

(b) Evolution of the RMSE along the scenario.

Fig. 6. Evolution of weights and $\sqrt{\text{MSE}}$ (RMSE) for our adaptive approach of Bernstein polynomial curve regression of degree 8, $\lambda = 0.9$, $\alpha = 0.08$. Scenario describes a nominal period between 2 periods of jamming. #39 corresponds to a link capacity of 11 Kbits/s, #10 40 Kbits/s and #1 49 Kbts/s. Figure is smoothed over 100 points.

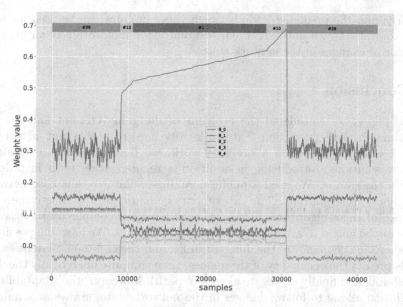

(a) Evolution of weights along the scenario.

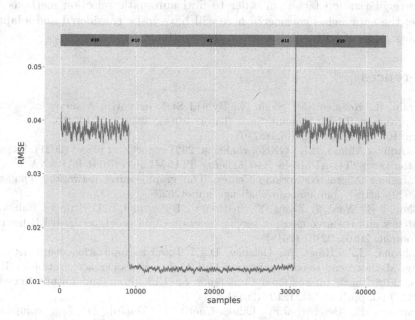

(b) Evolution of the RMSE along the scenario.

Fig. 7. Evolution of weights and $\sqrt{\text{MSE}}$ (RMSE) for our adaptive approach of Linear Regression, $\lambda = 0.9$, $\alpha = 0.08$. Scenario describes a nominal period between 2 periods of jamming. #39 corresponds to a link capacity of 11 Kbits/s, #10 40 Kbits/s and #1 49 Kbts/s. Figure is smoothed over 100 points.

regularly, and the predicted KPIs between these updates can be used for intelligent routing. As we have observed, residual error monitoring is an excellent indicator of changes in the network state.

6   Conclusion

In this paper, we considered the problem of designing efficient and low-cost algorithms for KPI prediction that are locally implementable and adaptive to network changes. Based on a previous work, we have argued and developed adaptive solutions, introducing in addition a regularization term in order to stabilize the results. We used a public domain simulator to simulate networks and generate relevant data. The experiments demonstrate the effectiveness and relevance of these algorithms. Thus, we now have low-complexity models that can be implemented iteratively at the level of local links. We have the possibility to predict the occupancy of the different links, and the end-to-end latencies (the models predict the occupancy of the queues, then compute analytically the delay for each link and finally aggregate along the path). Moreover, the adaptability of the solution allows to follow changes in the network state, always at a minimal cost, by re-adapting from the current solution and new data. The continuation of the work will focus on the choice criteria of the forgetting factor, on the impact of the regularization factor, in order to find automatic selection methods. Of course, the approaches considered here will have to be considered and adapted for other types of KPI, such as error rate or jitter.

References

1. Amin, R., Reisslein, M., Shah, N.: Hybrid SDN networks: A survey of existing approaches. IEEE Commun. Surv. Tutor. **20**(4), 3259–3306 (2018). https://doi. org/10.1109/COMST.2018.2837161
2. de Aquino Afonso, B.K.: GNNet challenge 2021 report (1st place) (2021). https:// github.com/ITU-AI-ML-in-5G-Challenge/ITU-ML5G-PS-001-PARANA
3. Barcelona Neural Networking Center: The graph neural networking challenge (2020). https://bnn.upc.edu/challenge/gnnet2020
4. Chua, F.C., Ward, J., Zhang, Y., Sharma, P., Huberman, B.A.: Stringer: Balancing latency and resource usage in service function chain provisioning. IEEE Internet Comput. **20**(6), 22–31 (2016)
5. Jahromi, H.Z., Hines, A., Delanev, D.T.: Towards application-aware networking: Ml-based end-to-end application KPI/QoE metrics characterization in SDN. In: 2018 Tenth International Conference on Ubiquitous and Future Networks (ICUFN), pp. 126–131. IEEE (2018)
6. Larrenie, P., Bercher, J.F., Lahsen-Cherif, I., Venard, O.: Low complexity approaches for end-to-end latency prediction. In: Proceedings of the 13th IEEE International Conference On Computing, Communication and Networking Technologies. IEEE (2022)
7. Pasca, S.T.V., Kodali, S.S.P., Kataoka, K.: AMPS: Application aware multipath flow routing using machine learning in SDN. In: 2017 Twenty-third National Conference on Communications (NCC), pp. 1–6. IEEE (2017)

8. Poularakis, K., Iosifidis, G., Tassiulas, L.: SDN-enabled tactical ad hoc networks: Extending programmable control to the edge. IEEE Commun. Magaz. **56**(7), 132–138 (2018)

9. Poularakis, K., Qin, Q., Nahum, E.M., Rio, M., Tassiulas, L.: Flexible SDN control in tactical ad hoc networks. Ad Hoc Netw. **85**, 71–80 (2019). https://doi.org/10.1016/j.adhoc.2018.10.012

10. Rusek, K., Suárez-Varela, J., Mestres, A., Barlet-Ros, P., Cabellos-Aparicio, A.: Unveiling the potential of graph neural networks for network modeling and optimization in SDN. In: Proceedings of the 2019 ACM Symposium on SDN Research, pp. 140–151 (2019)

11. Singh, S., Jha, R.K.: A survey on software defined networking: Architecture for next generation network. J. Netw. Syst. Manag. **25**(2), 321–374 (2017)

12. Suárez-Varela, J., et al.: The graph neural networking challenge: A worldwide competition for education in AI/ML for networks. ACM SIGCOMM Comput. Commun. Rev. **51**(3), 9–16 (2021). https://doi.org/10.1145/3477482.3477485

13. Varga, A., Hornig, R.: An overview of the omnet++ simulation environment. In: 1st International ICST Conference on Simulation Tools and Techniques for Communications, Networks and Systems (2010)

14. Widrow, B., Stearns, S.: Adaptive Signal Processing. In: Oppenheim, A.V. (ed.) Prentice-Hall (1985)

TDMA-Based MAC Protocols Designed or Optimized Using Artificial Intelligence for Safety Data Dissemination in Vehicular Ad-Hoc Network: A Survey

Maroua Ghamri[1], Selma Boumerdassi[2(✉)], and Aissa Belmeguenai[3]

[1] Electronic Department, University of Jijel, Jijel, Algeria
[2] CNAM, Paris, France
selma.boumerdassi@inria.fr
[3] Electronic Department, University of Skikda, Skikda, Algeria

Abstract. Vehicular Ad-hoc Network (VANET) has attracted the research community's attention in recent years due to its secure and safe applications on roads, ensuring unidentified transmission and advanced safety. An advanced safety system has been put in place to guarantee that all messages exchanged between vehicles are reliable. The Medium Access Control (MAC) protocols are needed to transmit the data between vehicles. There are different types of roads (highway, urban), then the design of these protocols changes according to the density of vehicles. The jam or the high mobility could cause a problem of collision to access the channel. To avoid these cases and to decrease hazards on roads (accidents), scientists have proposed many techniques to optimize the control of TDMA-based MAC protocols. Our object is to investigate the proposed MAC protocols and focus more on the TDMA-based protocols optimized by the need for Artificial Intelligence (state-of-art) to ensure safe and reliable IV (Inter-vehicles) communications. The classification, the Artificial Intelligent methods, and the proposed real datasets for optimizing these protocols are presented in this work.

Keywords: VANET · MAC layer classifications · TDMA · Artificial Intelligence

1 Introduction and Background

VANET or autonomous vehicles consist of a set of vehicles traveling along the roads that can communicate with each other via ad-hoc wireless devices. VANET is a subgroup of Mobile Ad-hoc Network (MANET), and the difference between them could be seen in the density and the mobility of nodes which are extremely high in VANET. The high level of connectivity on roads makes communication in VANET more difficult. For autonomous vehicles' communication in different road types (highway, urban, city) the three functions in OB systems must be used: the

É. Renault and P. Mühlethaler (Eds.): MLN 2022, LNCS 13767, pp. 88–112, 2023.
https://doi.org/10.1007/978-3-031-36183-8_7

Telemetry Elements: Radars, Lidars, Cameras, and Image Processing, The Geo-Positioning/Time: Global Positioning System (GPS), Global Navigation Satellite System (GNSS), GALILEO, and the Omnidirectional/Unidirectional antennas for different types of vehicles [1]. Every vehicle is occupied with OBU (On-board Unit) represented in GPS, camera, antenna, etc. There are different network models that are deployed to help vehicles communicate with each other in a good manner.

1.1 Vehicular Communications Types

1.1.1 Vehicle-to-Vehicle (V2V) Communications
Vehicles or OBUs could directly exchange messages between them in real-time, Three major components are essential in vehicular communication: The Dedicated Short Range (DSRC), GPS, and antennas. **DSRC** radio channel [2] is divided into Control channel CCH and SCHs [3] to transmit and receive the data. The guard intervals are needed for synchronization and switching between CCH and SCHs [3]. The CCH interval for safety messages dissemination and the other channels for exchanging non-safety messages [3]. **GPS Receiver** [2]: is responsible for calculating the timestamp, the position, the speed, the direction, and the acceleration of the vehicles that are receiving via antennas. **The antennas** [2] are used to propagate the DSRC and GPS signals and guaranteed the transmission.

1.1.2 Neighbors-to-Neighbors (N2N) Communications [1]
V2V communications generally use an omnidirectional antenna which could not be the optimum solution for canceling loose in telemetry calculations by OB system in platoons. In the same lane, collisions occur between two contiguous members. Therefore, unidirectional communications provided by two unidirectional antennas in the rear and front of the vehicle are satisfied for changing messages and beacons with short-range which represent N2N (neighbor-to-neighbor) communications. Cohort form of a platoon that uses this type of communication.

1.1.3 Vehicle-to-Everything (V2X)
In this communication, Vehicles could communicate with each other by the need of Cellular [5] and cloud technologies.

– Cellular Vehicular-to-Everything (C-V2X) communications

Cellular V2X is a type of V2X where long-range technologies are included such as 3G, 4G LTE, 5G, and 6G. These technologies could guarantee the transmission/reception of messages due to their reliable infrastructures. The connectivity between the user and the base station could be done via an interface [6]. The C-V2X is preferred for the transmission of periodic messages such as Cooperative Awareness messages (CAMs) and Basic safety messages (BSMs) [6]. The S-TDMA [6] protocol is using LTE-V2X to address safe transmission messages.

– Vehicular-to-Cloud (V2C) communications

To ensure the long range communication with low cost the internet-based vehicle-to-cloud communication [7] is used. The RSUs is not required in this type of communication [7] which makes the communication cheaper.

– Vehicle-to-Road side unit (V2R) Communications

RSUs which is an equipment or device integrated into the edge of roads to communicate with vehicles (OBUs) or to communicate with each other to ensure more safe messages exchanging. Recently, different types of servers are implemented in RSUs to ensure V2X communications, i.e. in [4] the MEC (Mobile Edge Computing) server is used to decrease collision in the TDMA-based MAC protocol design for safety data dissemination.

1.2 Safety Messages Types in Facilities/Application Layer

1.2.1 Cooperative Awareness Messages (CAMs)

CAMs messages under the EN 302 637-2 that are depending on all ITS stations are types of periodic messages in which the vehicles exchange their current position, speed, and direction periodically [8]. The transmission of this message every 100 ms [9] and under four changing parameters: 4 m in the distance, 4°C in the heading, and 0.5m/s in the speed.

1.2.2 Decentralized Environmental Notification Messages (DNMs)

DNMs are a type of message broadcasted only in specific conditions or to detect some problems in traffic to avoid dangerous situations and it is based on geographical areas addressing [10].

1.2.3 Local Dynamic Map Messages (LDM)

LDMs messages in ETSI TR 102 863 are used for exchanging information about the objects that influence traffic such as maps of vehicles [11,12].

1.2.4 Cooperative Perception Messages (CPM) [12,13]

On roads; there are many types of anomalies such as holes, distortions, and dangerous objects. The CPMs under ETSI TS 103 324 standardization are used to detect unsafe objects on roads. The generation of this type of message is for specific vehicles. The CPMs messages are divided into PDU (Protocol Data Unit), and 5 messages. PDU header supports only the protocol version, message ID, and Station ID, the other information (speed, width, height, acceleration,...etc) of vehicles are included in the other 5 contains.

The different types of messages could be exchanged between nodes by different types of MAC protocols.

1.3 Motivations

The dissemination of messages in a safe manner is a challenging task in autonomous vehicles. Due to high changes in the topology, the share of the channel in the medium access layer between nodes is some well difficult and doesn't guarantee a delay to access a channel. The design of the Media Access Control (MAC) protocols demand complex multiple access channel mechanisms such as TDMA, FDMA, CDMA, and CSMA. Therefore; there are many authors who propose MAC protocols under DSRC standardization such as IEEE.802.11p [14] protocol that utilizes Carrier Sense Multiple Access (CSMA/CA) to reduce accident by solving the problem of collisions. IEEE.802.11p suffer from the hidden collision.To reduce more types of collisions as well the cost, TDMA-based MAC protocols are more needed. This type of protocol has been used to avoid different collisions by using different types of optimization methods. In this paper, we will survey some design of TDMA-based MAC protocols that utilizes artificial intelligent algorithms or networks to enhance safety communication.

1.4 Contributions of This Paper

- Classification of different MAC protocols that have been optimized or enhanced their design using Artificial Intelligent (AI) methods.
- The different techniques of Artificial Intelligent that have been used to optimize MAC protocols and proposed others.
- Some real datasets that could be adopted using it to exchange in the MAC layer.

The paper is organized as given in Fig. 1.

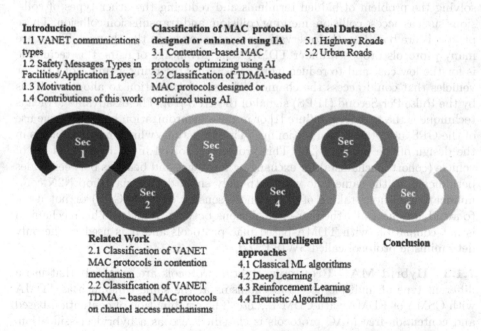

Introduction
1.1 VANET communications types
1.2 Safety Messages Types in Facilities/Application Layer
1.3 Motivation
1.4 Contributions of this work

Classification of MAC protocols designed or enhanced using IA
3.1 Contention-based MAC protocols optimizing using AI
3.2 Classification of TDMA-based MAC protocols designed or optimized using AI

Real Datasets
5.1 Highway Roads
5.2 Urban Roads

Related Work
2.1 Classification of VANET MAC protocols in contention mechanism
2.2 Classification of VANET TDMA – based MAC protocols on channel access mechanisms

Artificial Intelligent approaches
4.1 Classical ML algorithms
4.2 Deep Learning
4.3 Reinforcement Learning
4.4 Heuristic Algorithms

Conclusion

Fig. 1. Organization of the paper

2   Related Work

2.1   Classification of MAC Protocols in VANETs

In general, the MAC protocols for VANETs can be classified into three categories [15]: contention-based, contention-free, and hybrid Mac protocols.

2.1.1   Contention-Based MAC Protocols

The transmission in contention-based is based on the random transmission of the message after listing the free access channel which is represented on CSMA/ CA. The well-known protocol in this classification is IEEE.802.11p which gave a good performance to enhance the use of the channel efficiency due to different good modulations used in the physical layer of this protocol under the technology DSRC and the details are in [14]. This protocol suffers from hidden collision, especially in high density, in [16] where all types of collisions are well explained. The S-TDMA [6] is a contention-free MAC protocol, the most competitive to IEEE.802.11p, the Technical Committee Intelligent Transport (ETSI) [17] ensures and validates that the mechanism in this protocol could replace CSMA/CA [6]. In another survey [18], the protocols based on Additive Links On-line Hawaii Area (ALOHA) mechanism [19] are added as Contention-based protocols: ALOHA [20], S-ALOHA [21], MACA [22], MACAW [23], R-ALOHA [24], well-known RR-ALOHA+ [25].

2.1.2   Contention-Free MAC Protocols

In these protocols, the TDMA, CDMA, and FDMA mechanisms are used to guarantee fair access channels by solving the problem of hidden terminals and reducing the other types of collisions such as access collision, merging collision, and transmission collision. These protocols are based on the reservation to access or leave the channel. There are many protocols proposed under TDMA. The advantage of using this technique is for the low cost and to reduce more collision by reserving a time slot (Ts) for vehicles that could access the channel. The synchronization to allocate a Ts is by the Pulse Per Second (1PPS) signal of the GPS [15]. The disadvantage of this technique is the telemetry failure [1] or in the synchronization because of the use of the GPS and other Geo-Positioning /Time functions which make problems in the design of the protocol [26]. This problem is solved in [1] by type of set of vehicle (cohort mechanism) for exchanging messages and beacons between nodes periodically in the same lane with extremely small delays via 1-hop N2N communication without taking of care remote sensing devices (GPS) or not used, to avoid automatically the rear-end collisions between vehicles. This mechanism is not compatible with TDMA-based mac protocols and it is used in the only deterministic protocol called SWIFT [27].

2.1.3   Hybrid MAC Protocols

Hybrid protocols are protocols that use a different type of multiple access mechanisms at the same time such as TDMA with CSMA or FDMA with CDMA.....etc. The combination of contention-based and contention-free MAC protocols is classified also as a hybrid classification. The protocols are mentioned in [16,17]: CBRC [28], CS-TDMA [29], SOFT-MAC [30], DMMAC [31], HER-MAC [32], CBMMAC [33], CBMCS [34], R-MAC [35] and HTC-MAC [36].

2.2 Classification of VANET TDMA-Based MAC Protocols on Channel Access Mechanisms

This section presents the classification of contention-free TDMA – based MAC protocols: distributed, centralized and hybrid on single-channel and multi-channel. Most of these protocols are surveyed in [16], and we will add others.

2.2.1 Distributed Single-Channel Contention-Free TDMA – Based MAC Protocols
In the distributed MAC protocol, there is no centralization and the TDMA technique is used for the allocation of a time slot in the packet. A single channel means that the access channel is not halved to the data channel and to the control channel [37]. MS-ALOHA [38], RTOB [39] and eCAH-MAC [40] protocols like the first protocol VeMAC [41] are used the Frame Information (FI) message to determine the information of time slot status, the status of Ts could be busy, free, or in a collision to avoid a type of collisions. d-TiMAC [42] and ASMAC [43] are used in the four moving directions: north, east, west, and south to allocate a time slot in which this approach helps to reduce the number of collisions. In PT-MAC [44], the incremented technique is adopted for the allocation of Ts [16]. AS-DTMAC [45] protocol is the enhancement of DTMAC [46] added a random key to reserve a time slot.

2.2.2 Distributed Multi-channel Contention-Free TDMA-Based MAC Protocols
The limitation of the Single channel by using the same Ts to transmit the data that produce a problem and it isn't appropriate for DSRC multi-channel technology [41]. The most protocols are the enhancement of VeMAC [41] protocol such as A-VeMAC [47], VeMac-5 [48], Modified-VeMAC [49]. VeMAC is used the CCH and SCHs channels to reserve a time slot with the help of the RSU. ASTSMAC [26] protocol unlike the author protocols is used multiple frequencies with the same time slot to reduce more number of collisions.

2.2.3 Centralized Single-Channel Contention-Free TDMA-Based MAC Protocols
The reservation of a time slot for each vehicle is based on the RSUs or on the Cluster Head (CH) or both. The most Centralized single-channel protocols proposed are RSU-based approach which means the channel is dived into time slots, and the RSUs allocate a time slot to each vehicle in their safe coverage area. The protocols VAT-MAC [50] and CT-MAC [51] are examples.

2.2.4 Centralized Multi-channel Contention-Free TDMA-Based MAC Protocols
The approach of these protocols is based on the reservation of time slots by the control of RSUs and Cluster Head. The known Centralized Cluster-based protocols are the most used for multi-channel applications. Both SCHs and CCH channels are needed in which the cluster head is responsible for the management of time slots. The allocation of time slots is by the RSUs and also for the communication between clusters. The protocols TA-MAC [52] and TCM-MAC [53] are used in this approach.

3 Classification of Artificial Intelligent-Based MAC Protocols

In this section, we will classify the protocols that are used AI methods to enhance their design and are optimized into two classifications: Contention-based and Contention-free. The most contention-free MAC protocols that used AI are based on the TDMA mechanism. Figure 2 shows the classification of these protocols.

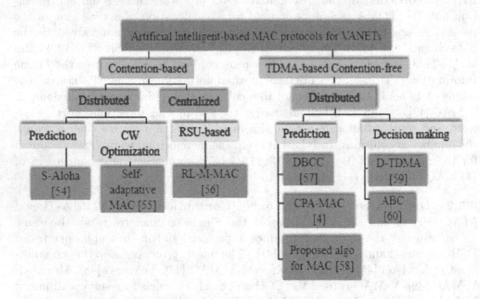

Fig. 2. The classification of MAC protocols designed or enhanced using AI

3.1 Contention-Based MAC Protocols Optimizing Using AI

The contention-based protocols suffer from hidden terminal or transmission collisions. Several types of artificial intelligent design optimizing methods are used to reduce this type of collision.

– S-ALOHA

The S-ALOHA protocol divided the channel into time slots with fixed duration in which the nodes could transmit data at the beginning of the time slot in a random way. When one vehicle Vx could access one-time slot Ts is considered a successful transmission, and the percentage of this happening is 37%. The authors in [54], are used supervised time-series prediction-based deep neural network algorithms (The Long Short-term memory (LSTM) and transformers) to predict the correct number of vehicles in the coverage area of each Base Station (BS) and in the duration of one second of the Luxemburg City even when the inputs are noisy. C-V2X communications such as 3G, 4G, and 5G are used in this protocol (Fig. 3).

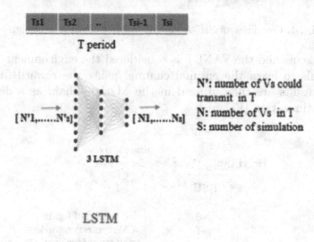

Fig. 3. Overview of S-Aloha Protocol Design

– Self-adaptative MAC

The Self-adaptative MAC protocol is used supervised deep reinforcement learning (Deep Q Network) to optimize the contention window (CW) for safety message exchanging. The contention window represents the maximum duration of nodes that could be waiting before transmission. In this network as seen in the figure, the vehicles act as agents in the vehicular environment. The agents learn the optimal CW by taking the transmitted, and received packets, and using the reward function r(t). This protocol could be classified as the only one that could present the full distribution (no infrastructures, no RSU) (Fig. 4).

– RL-M-MAC

RL-M-MAC protocol is used unsupervised reinforcement learning based on Bernoulli logic to avoid collisions by optimizing the contention window. The

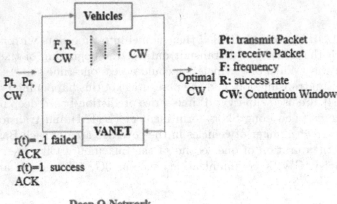

Deep Q-Network

Fig. 4. Overview of Self-Adaptive MAC Protocol Design

RSUs act as agents and the VANET is considered the environment. The agents take the signals to learn the optimal channel using the reward function r(t). Some of the results are then utilized in the Markov chain as a deep learning method for optimization (Fig. 5).

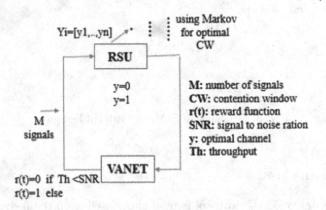

Deep Reinforcement Learning

Fig. 5. Overview of RL-M-MAC Protocol Design

3.2 Classification of TDMA-Based Contention-Free MAC Protocols Designed or Optimized Using AI

3.2.1 Prediction Distributed TDMA-Based MAC Protocols The distributed time slot reservation means the vehicles form one-hop neighbors (OHNs) and two-hop neighbors (THNs) to communicate with each other and the vehicles in the THNs should allocate distinct Ts.

– DBCC

There are three steps to designing the Distributed Beacon Congestion Control protocol. In the first step; the line conditions are predicted by using supervised machine learning algorithms (Naïve Bayes (NB) and Support Vector Machine (SVM)). The features or the inputs to learn the data are the history of Packet Delivery Ration (PDRs). The vehicles transmit periodically in each 100 ms: the sequence number, the latitude, the longitude, the altitude, and the speed. The output or the target of this protocol for fairness design are the NLos (Non-line-of-sight) and Los (Line-of-sight) conditions. When two vehicles see each r the NLos—0 and Los =1 (represent good probability = 97%). To check the results here; Accuracy, Precision, Recall, and False positive rate metrics are verified. Secondly; for distributed congestion control, every vehicle broadcast the link weight information (calculation of PDR, and link conditions) of its and its neighbors to satisfy the fairness of each frame. The length of the frame is 100ms which is divided into slots in each THNs range. The beacon rate is trapped 1 Hz 10 Hz (equal to 1 to 10 beacon/frame). To decrease or avoid transmission collisions, the beacon rate must be under the total number of slots that the frame support. Finally, greedy heuristic algorithms are used to optimize and give a beacon rate to each vehicle (Fig. 6).

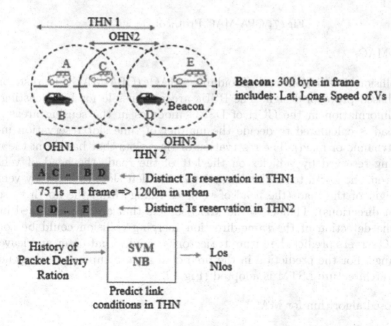

Fig. 6. DBCC MAC Protocol Design

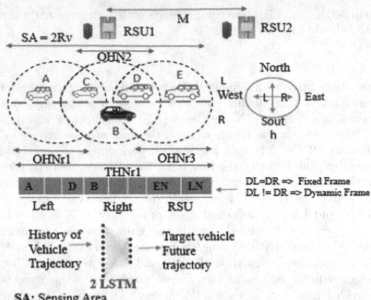

SA: Sensing Area
Rv: vehicles transmission range
L: left direction
R: right direction
DL: density of vehicles in left direction
DR: density of vehicles in right direction

Fig. 7. CPA-MAC Protocol Design

– CPA-MAC

The Collision Prediction and Avoidance MAC (CPA-MAC) is distributed TDMA-based protocol design. The RSUs are responsible for broadcasting the control information in the CCH of DSRC modules in the sensing areas. The traffic load is calculated to decide the manner of time slot reservation in the frame (dynamic or fixed). The reservation of time slots that based on: the right slots being reserved by vehicles on the left of the road (the head of vehicles change from the north to the left directions), and left slots reserved by vehicles on the right of the road (the head of vehicles change from the north to south and right directions). Thanks to the MEC servers that are implemented in the RSUs, the detection of the same-direction merging collision could be solved. The MEC servers predict the future trajectory of nodes that enter and leave the RSUs range. For the prediction in the predicted area, a supervised deep neural network architecture LSTM is adopted (Fig. 7).

– Proposed algorithm for MAC

To realize MAC protocol in C-V2X communication (5G technology), the data and the real-time load balancing scheduling are transmitted in the spectrum of cellular 5G communication. The intelligent traffic prediction model used here is supervised deep learning architecture (CNN). The algorithm is compared with algorithms of another TDMA-based multichannel MAC protocol and it presented very good performance. Therefore; this algorithm must be adopted to design a TDMA-based Multi-channel MAC protocol to test their performance. The algorithm solves the problem of collision in cooperative transmission by the use of a heterogeneous network as seen in Fig. 8 (Table 1).

3.2.2   Decision Making Distributed TDMA-Based MAC Protocol

– D-TDMA

Distributed-TDMA protocol is used deep learning techniques to make a decision of keeping a reservation of the same time slot or releasing it in the THNs frame. The differentiation Markov chain method is used to know if the channel is busy or not busy. There are two states for this method. When the vehicles could estimate the collision, they release a time slot considered as a good state. And if the nodes couldn't estimate the collision, it released the time slot and was considered as a bad state. The vehicles used the speed and position information to know the channel state. The nodes follow a Poisson distribution and their speeds follow the truncated normal distribution (Fig. 9).

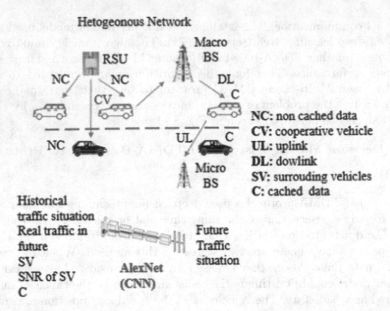

Fig. 8. Proposed Predicted Algorithm for Distributed TDMA-based MAC Protocol Design

– ABC

In the ABC (Adaptive Beacon Control) protocol, the vehicles exchange the beacons periodically. Due to the limited bandwidth of DSRC radio in high density, the transmission of the beacon rate of messages must be controlled for fair communication. The beacon message contains the beacon rate and the danger coefficient. The danger coefficient parameter depends on the velocity, acceleration, deceleration, and distance. To solve rear-end collision between neighboring vehicles this value must be big when the velocity increase and distance decreases and vice versa. In distributed TDMA, the duration of the frame is divided into fixed time slots and the vehicles Vi must send the beacon rate lower than S duration. If the reservation of Ts of Vi in the same THNs is not fair, a collision may occur. So, the greedy heuristic algorithm is used to make decisions for using Ts to each Vi (Fig. 10 and Table 2).

4 Artificial Intelligent MAC Protocols Optimizing Methods

In this section, we will summarize the different AI methods to design the MAC protocols (see Fig. 11).

4.1   Classical Machine Learning Algorithms

4.1.1   Supervised Learning Supervised learning is a method of AI where the data is labeled which means the output is specified when training the inputs. The most supervised machine learning algorithms are used to optimize the TDMA prediction-based MAC protocols. The algorithms SVM and NB are used to predict line conditions before transmission in DBCC to reduce the number of collisions. The K-Nearest Neighbor (KNN) algorithm is very used in VANET to predict the time and space road traffic and this algorithm gives good results in [75]. The principle of this algorithm is to use feature similarity to predict the new point in data such as time, space, and positions of vehicles.

4.1.2   Unsupervised Learning Unsupervised learning is a method of IA where the data is unlabelled which means the clusters obtained by training data. The principle of this algorithm is clustering input data depending on a different type of distance between points. The most clustering algorithms are existing in [76] and the most knowing and used algorithm is K-means. The K-means algorithm is used for clustering in real HighD dataset in [77,78] and it isn't gave good results compared with deep learning methods. The classical methods couldn't work well with a big dataset and it isn't preferred to verify the performance of Cluster-based MAC protocols.

4.2   Deep Learning

4.2.1   Supervised Learning The neural networks CNN, Graph Neural Network (GNN), and LSTM extension of Recurrent Neural Network (RNN), and RNN models are the most used for time-series prediction in a vehicular network. The CNN architecture is used in the MF-TDMA [79] protocol to predict the spectrum information of neighbors in wireless networks (general network, not VANET). The RNN with LSTM architectures in [12] is combined to predict the transmission of neighbors and to have safe messages exchanged in V2X communication. In [58], the three CNN, RNN, and LSTM architectures are compared

Table 1. Comparison between Prediction Distributed TDMA-based MAC protocols optimizing using AI

Protocol	DBCC	CPA-MAC	Proposed algo for MAC
Year	2018	2021	2021
Communication types	V2V	V2V, V2R	C-V2X, V2R
Channel	Multichannel	Multichannel	/
Road type	Highway, Suburban, Urban	Highway	Road in City
Collision avoidance	Rear-end Collision	Merging Collision	Access Collision
AI Type	Supervised Machine Learning	Supervised Deep learning	Supervised Deep Learning
AI algorithms or architectures	NB, SVM algorithms	LSTM architecture	CNN architecture
Dataset	3 dataset in Shanghai city	NGSIM dataset [62]	Xueyuan road dataset, Beijing city, in May to August, 2020
Messages types	BSM (Basic Safety messages)	/	/
Hopes count	Two	Two	/
Compared with	Conventional 802.11p	VeMAC, VeMAC-5, A-VeMAC	MC-MAC [66], sdnMAC [67], VeMAC [68], S-TDMA [69], ADHOC-MAC [70]
Simulators	MATLAB [61]	OMNET++ [63], SUMO [64]	Python 3.6.1 [71], SUMO
Performance metrics	Rate of reception collision	Throughput Rate of collision Packet Loss Rate Channel utilization	Throughput, Access Collision Rate, Throughput, Access time
Limitation	No guaranteed of the protocol design optimization	No communication between vehicles out of RSU coverage area	The number of channels and hop counts are not described and the errors of processing the data
Solution	Keep the time slot redundancy and optimize the design	Fog computing [65] to Vs communicate out of RSU coverage area	Multichannel MAC design with distributed time slot reservation is more preferred

to find the best one that could predict the road network under C-V2X communication. They find that the LSTM architectures give the best results in Accuracy, precision, Recall, and F1. The LSTM method could give good results in the prediction of a big and real dataset NCGIM and the details are in [80]. The LSTM architecture also shows their performance to predict the next time points in the traffic in [81]. The CPA-MAC protocol is used the LSTM architecture of [80] to prove their performance for safety data dissemination in TDMA-based MAC protocol. The GNN in [82] is used for traffic prediction in the HighD [83] and NCGIM datasets. To decide the Ts reservation of vehicles the Markov chain is used in the D-TDMA protocol. The Markov chain is a stochastic deep learning method to give a probability to each state.

4.2.2   Unsupervised Learning The GNN and CNN models are used for clustering for autonomous vehicles. The GNN architecture in [84] achieved good results for clustering vehicles in the HighD dataset with 0.97 accuracy and better results than the accuracy of using CNN architecture in [85].

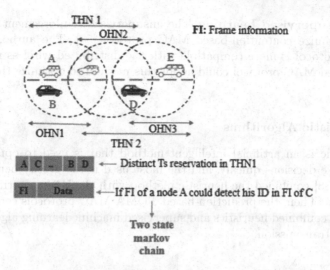

Fig. 9. Illustration of D-TDMA MAC Protocol Design

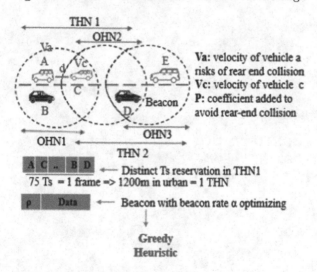

Fig. 10. ABC MAC Protocol Design

4.3 Reinforcement Learning

4.3.1 Supervised Learning Reinforcement learning is an artificial method that learned the optimal behavior of vehicles by observation of the environmental behavior. To optimize or make decisions in any network performance, reinforcement learning [86] is the most utilized. The contention-based MAC protocols use this learning method to optimize the CW. This architecture is also used to optimize the time scheduling for avoiding collisions in TDMA-based MAC protocol in [87]. The time-slot learning [87] is used in general wireless networks, not for VANET, but it could give a good performance in VANET.

4.3.2   Unsupervised Learning The unsupervised reinforcement learning is used to optimize contention-based MAC protocol [55]. The authors declared that this protocol is more compatible with the distributed wireless network. A TDMA-based MAC protocol could adopt this method to optimize the time slot scheduling.

4.4   Heuristic Algorithms

The heuristic is an artificial intelligent method that is used to optimize, predict, or make decisions quickly, and the most used is the greedy heuristic. The authors in [88] combined the heuristic method with the KNN algorithm for the optimization. Then, the prediction-based TDMA MAC protocols could be optimized using combined heuristics and supervised machine learning algorithms for safety data transmission.

Table 2. Comparison between Making Decision Distributed TDMA-based MAC protocols optimizing using AI

Protocol	D-TDMA	ABC
Year	2017	2018
Communication types	V2V	V2V , V2R
Channel	Single channel	Single channel
Road type	Highway	Urban
Collision avoidance	transmission Collision	Rear-end Collision
AI Type	Supervised deep Learning	Artificial Intelligent method
AI algorithms or architectures	Markov chain	Greedy heuristic algorithms
Dataset	Highway 401 of the Canadian province of Ontario [72]	/
Messages types	/	Beacons
Hopes count	Two	Two
Compared with	/	Conventional 802.11p, LIMERIC [74]
Simulators	MATLAB, PTV Vissim [73]	Python, SUMO
Performance metrics	transmission collision Rate	beacon rate transmission and reception, efficiency, collision rate
Limitation	the single channel is not preferred for vehicles communication, if the vehicles release the time slot they could collide	the single channel and heuristic algorithm are not preferred
Solution	Reinforcement learning is more efficient to make decision as IA method	Using Reinforcement learning

5   Real Dataset

The road type affects the MAC protocol's performance. The density of vehicles in urban is higher than on highways which means the reservation of time slots changed in each area. In this section, we will give some datasets that gave good results in different machine learning models.

5.1   Highway Roads

– HighD dataset

The HighD dataset is an open-source proposed dataset for safe autonomous vehicles application. Until 2020; this dataset is used in communication such as estimation of the CAM messages in [9] and in clustering in [77,78]. This dataset gives a lot of information for tracking 110 500 vehicles and is the most suited to verify the TDMA-based MAC protocol design optimization.

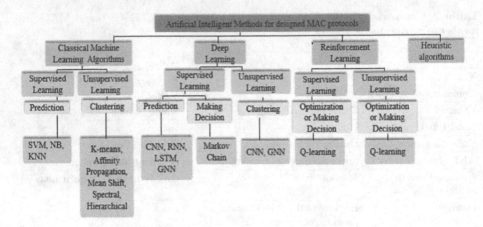

Fig. 11. Artificial Intelligent Methods for Enhancing MAC Layer Protocol Designing

- CAM dataset [89]

The Dataset of Cooperative Awareness Messages (CAM) is France's C-ITS project which contains 10,174,437 CAM messages. This dataset isn't available yet but it is the most needed for exchanging real data dissemination in the MAC layer.

5.2   Urban Roads

- Argoverse dataset [90]

Agroverse consists of two released open-source datasets in US cities, the first gives details of 3D maps and 324557 tracking vehicles in two cities Miami and Pittsburgh. The second release contains four vehicle tracking datasets of six cities Austin, Detroit, Miami, Pittsburgh, Palo Alto, and Washington, D.C.

6   Conclusion

There are two types of TDMA-based MAC protocols: Centralized and Distributed. There are only four TDMA-based MAC protocols and one algorithm to design it using AI proposed in VANETs. The Predicted Distributed MAC protocols are most proposed and needed in VANETs. These protocols are mostly designed using LSTM models and it is the most preferred. The Markov chain and greedy heuristic algorithms are used to make decisions for time slot reservations. Deep Reinforcement Learning is more needed to optimize and make decisions on MAC protocol. Classical Machine Learning algorithms are not preferred to model in Big Datasets or to design MAC protocol. Deep learning methods are the most needed due to the large information in data. The Datasets are an essential part to design MAC protocol using AI in urban or on highways. The CAM dataset could be used to model TDMA-based MAC protocol on highway roads. The Agroverse datasets could be used to model TDMA-based MAC protocol in the urban road. As a future work, we will design the first centralized TDMA-based MAC protocol using the proposed deep learning methods and datasets.

References

1. Le Lann, G.: Cohorts and groups for safe and efficient autonomous driving on highways. In: 2011 IEEE Vehicular Networking Conference (VNC), pp. 1–8. IEEE (2011). https://doi.org/10.1109/VNC.2011.6117117
2. Demba, A., Möller, D.P.: Vehicle-to-vehicle communication technology. In: 2018 IEEE International Conference on Electro/Information Technology (EIT), pp. 0459–0464. IEEE (2018). https://doi.org/10.1109/EIT.2018.8500189
3. Chen, Q., Jiang, D., Delgrossi, L.: IEEE 1609.4 DSRC multi-channel operations and its implications on vehicle safety communications. In: 2009 IEEE Vehicular Networking Conference (VNC), pp. 1–8. IEEE (2009). https://doi.org/10.1109/VNC.2009.5416394
4. Liu, B., et al.: Cpa-mac: A collision prediction and avoidance mac for safety message dissemination in mec-assisted vanets. IEEE Trans. Netw. Sci. Eng. 9(2), 783–794 (2021)
5. Gyawali, S., Xu, S., Qian, Y., Hu, R.Q.: Challenges and solutions for cellular based V2X communications. IEEE Commun. Surv. Tutor. 23(1), 222–255 (2020)
6. Gallo, L., Härri, J.: Analysis of a S-TDMA distributed scheduler for ad-hoc cellular-V2X communication. Ad Hoc Netw. 88, 160–171 (2019)
7. Sliwa, B., Falkenberg, R., Liebig, T., Piatkowski, N., Wietfeld, C.: Boosting vehicle-to-cloud communication by machine learning-enabled context prediction. IEEE Trans. Intell. Transp. Syst. 21(8), 3497–3512 (2019)
8. Lyamin, N., Vinel, A., Jonsson, M., Bellalta, B.: Cooperative awareness in VANETs: On ETSI EN 302 637-2 performance. IEEE Trans. Vehicul. Technol. 67(1), 17–28 (2017)
9. Jaarsveld, F.: Estimating the effects of the Macroscopic Traffic Parameters on the overall Cooperative Awareness Message generations (Bachelor's thesis, University of Twente) (2020)
10. Ségarra, G.: Road co-operative systems-societal and business values. In: 2009 9th International Conference on Intelligent Transport Systems Telecommunications, (ITST), pp. 610–615. IEEE (2009). https://doi.org/10.1109/ITST.2009.5399282
11. ETSI TR 102 863 V1.1.1: Technical Report Intelligent Transport Systems (ITS); Vehicular Communications; Basic Set of Applications; Local Dynamic Map (LDM); Rationale for and guidance on standardization (2011)
12. Khan, M.I., Aubet, F.X., Pahl, M.O., Härri, J.: Deep learning-aided application scheduler for vehicular safety communication. arXiv preprint arXiv:1901.08872 (2019)
13. Thandavarayan, G., Sepulcre, M., Gozalvez, J.: Generation of cooperative perception messages for connected and automated vehicles. IEEE Trans. Vehicul. Technol. 69(12), 16336–16341 (2020)
14. Arena, F., Pau, G., Severino, A.: A review on IEEE 802.11 p for intelligent transportation systems. J. Sens. Actuat. Netw. 9(2), 22 (2020)
15. Hadded, M., Mühlethaler, P., Laouiti, A., Zagrouba, R., Saidane, L.A.: TDMA-based MAC protocols for vehicular ad hoc networks: A survey, qualitative analysis, and open research issues. IEEE Commun. Surv. Tutor. 17(4), 2461–2492 (2015)

16. Johari, S., Krishna, M.B.: TDMA based contention-free MAC protocols for vehicular ad hoc networks: A survey. Vehicul. Commun. **28**, 100308 (2021)
17. Final draft ETSI EN 302 636–1 V1.2.1 (2014). https://www.etsi.org. Accessed 1 July 2022
18. Hota, L., Nayak, B.P., Kumar, A., Ali, G.M.N., Chong, P.H.J.: An analysis on contemporary MAC layer protocols in vehicular networks: State-of-the-art and future directions. Future Internet **13**(11), 287 (2021)
19. What is ALOHA. https://ecomputernotes.com/computernetworkingnotes/communicationnetworks/what-is-aloha
20. Karn, P.: MACA-a new channel access method for packet radio. In: ARRL/CRRL Amateur Radio 9th Computer Networking Conference, vol. 140, pp. 134–140 (1990)
21. Bharghavan, V., Demers, A., Shenker, S., Zhang, L.: MACAW: A media access protocol for wireless LAN's. ACM SIGCOMM Comput. Commun. Rev. **24**(4), 212–225 (1994)
22. Tobagi, A.F., Kleinrock, L.: Packet switching in radio channels: Part II-the hidden terminal problem in carrier sense multiple-access and the busy-tone solution. IEEE Trans. Commun. 1417–1433 (1975)
23. Jin, T.K., Cho, H.D.: Multi-code MAC for multi-hop wireless Ad hoc networks. IEEE Vehicul. Technol. Conf. **2**, 1100–1104 (2002). https://doi.org/10.1109/VETECF.2002.1040774
24. Asadallahi, S., Refai, H.H.: Modified r-aloha: Broadcast mac protocol for vehicular ad hoc networks. In: 2012 8th International Wireless Communications and Mobile Computing Conference (IWCMC), pp. 734–738. IEEE (2012). https://doi.org/10.1109/IWCMC.2012.6314295
25. Cozzetti, H.A., Scopigno, R.: RR-Aloha+: A slotted and distributed MAC protocol for vehicular communications. In: 2009 IEEE Vehicular Networking Conference (VNC), pp. 1–8. IEEE (2009). https://doi.org/10.1109/VNC.2009.5416375
26. Li, S., Liu, Y., Wang, J.: ASTSMAC: Application suitable time-slot sharing MAC protocol for vehicular ad hoc networks. IEEE Access **7**, 118077–118087 (2019)
27. Le Lann, G.: Safe automated driving on highways-beyond today's connected autonomous vehicles. In: 8th Complex Systems Design & Management Conference "Towards smarter and more autonomous systems" (2017)
28. Tomar, R.S., Verma, S., Tomar, G.S.: Cluster based RSU centric channel access for VANETs. In: Gavrilova, M.L., Tan, C.J.K. (eds.) Transactions on Computational Science XVII. LNCS, vol. 7420, pp. 150–171. Springer, Heidelberg (2013). https://doi.org/10.1007/978-3-642-35840-1_8
29. Zhang, L., Liu, Z., Zou, R., Guo, J., Liu, Y. : A scalable CSMA and self-organizing TDMA MAC for IEEE 802.11 p/1609. x in VANETs. Wirel. Personal Commun. **74**(4), 1197–1212 (2014)
30. Abdalla, G.M., Abu-Rgheff, M.A., Senouci, S.M.: Space-orthogonal frequency-time medium access control (SOFT MAC) for VANET. In: 2009 Global Information Infrastructure Symposium, pp. 1–8. IEEE (2009)
31. Lu, N., Ji, Y., Liu, F., Wang, X.: A dedicated multi-channel MAC protocol design for VANET with adaptive broadcasting. In: 2010 IEEE Wireless Communication and Networking Conference, pp. 1–6. IEEE (2010)
32. Dang, D.N.M., Dang, H.N., Nguyen, V., Htike, Z., Hong, C.S. (2014). HER-MAC: A hybrid efficient and reliable MAC for vehicular ad hoc networks. In: 2014 IEEE 28th International Conference on Advanced Information Networking and Applications, pp. 186–193, IEEE (2014). https://doi.org/10.1109/AINA.2014.27

33. Su, H., Zhang, X.: Clustering-based multichannel MAC protocols for QoS provisionings over vehicular ad hoc networks. IEEE Trans. Vehicul. Technol. **56**(6), 3309–3323 (2007)
34. Ding, R., Zeng, Q.A.: A clustering-based multi-channel vehicle-to-vehicle (V2V) communication system. In: 2009 First International Conference on Ubiquitous and Future Networks, pp. 83–88, IEEE (2009). https://doi.org/10.1109/ICUFN.2009. 5174290
35. Guo, W., Huang, L., Chen, L., Xu, H., Miao, C.: R-mac: Risk-aware dynamic mac protocol for vehicular cooperative collision avoidance system. Int. J. Distrib. Sens. Netw. **9**(5), 686713 (2013)
36. Nguyen, V., Pham, C., Oo, T.Z., Tran, N.H., Huh, E.N., Hong, C.S.: MAC protocols with dynamic interval schemes for VANETs. Vehicul. Commun. **15**, 40–62 (2019)
37. Razfar, M., Abedi, A.: Single channel versus multichannel MAC protocols for mobile ad hoc networks. Proc. World Congr. Eng. Comput. Sci. **2**, 19–21 (2011)
38. Scopigno, R., Cozzetti, H.A.: Mobile slotted aloha for vanets. In: 2009 IEEE 70th Vehicular Technology Conference Fall, pp. 1–5. IEEE (2009). https://doi.org/10. 1109/VETECF.2009.5378792
39. Han, F., Miyamoto, D., Wakahara, Y.: RTOB: A TDMA-based MAC protocol to achieve high reliability of one-hop broadcast in VANET. In: 2015 IEEE International Conference on Pervasive Computing and Communication Workshops (PerCom Workshops), pp. 87–92. IEEE (2015). https://doi.org/10.1109/PERCOMW. 2015.7133999
40. Bharati, S., Zhuang, W., Thanayankizil, L.V., Bai, F.: Link-layer cooperation based on distributed TDMA MAC for vehicular networks. IEEE Trans. Vehicul. Technol. **66**(7), 6415–6427 (2016)
41. Omar, H.A., Zhuang, W., Li, L.: VeMAC: A TDMA-based MAC protocol for reliable broadcast in VANETs. IEEE Trans. Mob. Comput. **12**(9), 1724–1736 (2012)
42. Dragonas, V., Oikonomou, K., Giannakis, K., Stavrakakis, I.: A disjoint frame topology-independent TDMA MAC policy for safety applications in vehicular networks. Ad Hoc Netw. **79**, 43–52 (2018)
43. Han, S.Y., Zhang, C.Y.: ASMAC: An adaptive slot access MAC protocol in distributed VANET. Electronics **11**(7), 1145 (2022)
44. Jiang, X., Du, D.H.: PTMAC: A prediction-based TDMA MAC protocol for reducing packet collisions in VANET. IEEE Trans. Vehicul. Technol. **65**(11), 9209–9223 (2016)
45. Boukhalfa, F., Hadded, M., Muhlethaler, P., Shagdar, O. : An active signaling mechanism to reduce access collisions in a distributed TDMA based MAC protocol for vehicular networks. In: International Conference on Advanced Information Networking and Applications, pp. 286–300. Springer, Cham (2019). https://doi. org/10.1007/978-3-030-15032-7_25
46. Hadded, M., Laouiti, A., Mühlethaler, P., Saidane, L. A. : An infrastructure-free slot assignment algorithm for reliable broadcast of periodic messages in vehicular ad hoc networks. In: 2016 IEEE 84th Vehicular Technology Conference (VTC-Fall), pp. 1–7. IEEE (2016). https://doi.org/10.1109/VTCFall.2016.7880903
47. Chen, P., Zheng, J., Wu, Y.: A-VeMAC: An adaptive vehicular MAC protocol for vehicular ad hoc networks. In: 2017 IEEE International Conference on Communications (ICC), pp. 1–6. IEEE (2017). https://doi.org/10.1109/ICC.2017.7997358

48. Omar, H.A., Zhuang, W., Li, L.: Evaluation of VeMAC for V2V and V2R communications under unbalanced vehicle traffic. In: 2012 IEEE Vehicular Technology Conference (VTC Fall), pp. 1–5. IEEE (2012). https://doi.org/10.1109/VTCFall.2012.6398905
49. Kawakami, T., Kamakura, K.: Modified TDMA-based MAC protocol for vehicular ad hoc networks. In: 2015 IEEE International Conference on Pervasive Computing and Communication Workshops (PerCom Workshops), pp. 93–98. IEEE (2015). https://doi.org/10.1109/PERCOMW.2015.7134000
50. Cao, S., Lee, V.C.: A novel adaptive TDMA-based MAC protocol for VANETs. IEEE Commun. Lett. **22**(3), 614–617 (2017)
51. Wang, Y., Shi, J., Chen, L., Lu, B., Yang, Q.: A novel capture-aware TDMA-based MAC protocol for safety messages broadcast in vehicular ad hoc networks. IEEE Access **7**, 116542–116554 (2019)
52. El Joubari, O., Othman, J.B., Vèque, V.: TA-TDMA: A traffic aware TDMA MAC protocol for safety applications in VANET. In: 2021 IEEE Symposium on Computers and Communications (ISCC), pp. 1–8. IEEE (2021)
53. Pal, R., Prakash, A., Tripathi, R.: Triggered CCHI multichannel MAC protocol for vehicular ad hoc networks. Vehicul. Commun. **12**, 14–22 (2018)
54. Romo-Montiel, E., Menchaca-Mendez, R., Rivero-Angeles, M.E., Menchaca-Mendez, R.: Improving communication protocols in smart cities with transformers. ICT Express (2022)
55. Choe, C., Choi, J., Ahn, J., Park, D., Ahn, S.: Multiple channel access using deep reinforcement learning for congested vehicular networks. In: 2020 IEEE 91st Vehicular Technology Conference (VTC2020-Spring), pp. 1–6. IEEE (2020). https://doi.org/10.1109/VTC2020-Spring48590.2020.9128853
56. Kannan, K., Devaraju, M.: QoS supported adaptive and multichannel MAC protocol in vehicular ad-hoc network. Cluster Comput. **22**(2), 3325–3337 (2019)
57. Lyu, F., Cheng, N., Zhou, H., Xu, W., Shi, W., Chen, J., Li, M.: DBCC: Leveraging link perception for distributed beacon congestion control in VANETs. IEEE Internet Things J. **5**(6), 4237–4249 (2018)
58. Yu, M. : Construction of regional intelligent transportation system in smart city road network via 5G network. IEEE Trans. Intell. Transp. Syst. (2022)
59. Bharati, S., Omar, H.A., Zhuang, W.: Enhancing transmission collision detection for distributed TDMA in vehicular networks. ACM Trans. Multim. Comput. Commun. Appl. (TOMM) **13**(3s), 1–21 (2017)
60. Lyu, F., et al.: ABC: Adaptive beacon control for rear-end collision avoidance in VANETs. In: 2018 15th Annual IEEE International Conference on Sensing, Communication, and Networking (SECON), pp. 1–9. IEEE (2018)
61. Matlab. https://fr.mathworks.com/products/matlab.html. Accessed 3 July 2022
62. NGISIM Dataset. https://ops.fhwa.dot.gov/trafficanalysistools/ngsim.html
63. Andras Varga (Opensim Ltd) OMNET++, https://link.springer.com/chapter/10.1007/978-3-642-12331-3_3. Accessed 2 July 2022
64. Simulator for Urban Mobility. https://sumo.dlr.de/docs/index.html. Accessed 3 July 2022
65. Liu, K., Xu, X., Chen, M., Liu, B., Wu, L., Lee, V.C.: A hierarchical architecture for the future internet of vehicles. IEEE Commun. Magaz. **57**(7), 41–47 (2019)
66. Nguyen, V., Khanh, T.T., Oo, T.Z., Tran, N.H., Huh, E.N., Hong, C.S.: A cooperative and reliable RSU-assisted IEEE 802.11 p-based multi-channel MAC protocol for VANETs. IEEE Access **7**, 107576–107590 (2019)

67. Yang, T., Kong, L., Zhao, N., Sun, R.: Efficient energy and delay tradeoff for vessel communications in SDN based maritime wireless networks. IEEE Trans. Intell. Transp. Syst. **22**(6), 3800–3812 (2021)
68. Sun, Y., Kuai, R., Li, X., Tang, W.: Latency performance analysis for safety-related information broadcasting in VeMAC. Trans. Emerg. Telecommun. Technol. **31**(5), e3751 (2020)
69. Luo, G., Li, J., Zhang, L., Yuan, Q., Liu, Z., Yang, F.: sdnMAC: A software-defined network inspired MAC protocol for cooperative safety in VANETs. IEEE Trans. Intell. Transp. Syst. **19**(6), 2011–2024 (2018)
70. Lyu, F., et al.: Fine-grained TDMA MAC design toward ultra-reliable broadcast for autonomous driving. IEEE Wirel. Commun. **26**(4), 46–53 (2019)
71. Python 3.6.1. https://www.python.org/downloads/release/python-361/. Accessed 4 July 2022
72. Canadian Dataset. https://open.canada.ca/data/en/dataset/d12f5685-8ed9-486e-97de7a0ab72ef56d?wbdisable=true. Accessed 5 July 2022
73. PTV-VISSIM Simulator. https://www.ptvgroup.com/fr/solutions/produits/ptv-vissim/
74. Bansal, G., Kenney, J.B., Rohrs, C.E.: LIMERIC: A linear adaptive message rate algorithm for DSRC congestion control. IEEE Trans. Vehicul. Technol. **62**(9), 4182–4197 (2013)
75. Liu, Y., Yu, H., Fang, H.: Application of KNN prediction model in urban traffic flow prediction. In: 2021 5th Asian Conference on Artificial Intelligence Technology (ACAIT), pp. 389–392. IEEE (2021). https://doi.org/10.1109/ACAIT53529.2021.9731348
76. Clustering. https://scikit-learn.org/stable/modules/clustering.html. Accessed 4 Oct 2022
77. Hu, H., Lee, M.J.: Graph neural network-based clustering enhancement in VANET for cooperative driving. In: 2022 International Conference on Artificial Intelligence in Information and Communication (ICAIIC), pp. 162–167. IEEE (2022). https://doi.org/10.1109/ICAIIC54071.2022.9722625
78. Balasubramanian, L., Wurst, J., Botsch, M., Deng, K.: Traffic scenario clustering by iterative optimisation of self-supervised networks using a random forest activation pattern similarity. In: 2021 IEEE Intelligent Vehicles Symposium (IV), pp. 682–689. IEEE (2021)
79. Mennes, R., Claeys, M., De Figueiredo, F.A., Jabandžić, I., Moerman, I., Latré, S.: Deep learning-based spectrum prediction collision avoidance for hybrid wireless environments. IEEE Access **7**, 45818–45830 (2019)
80. Xu, W., et al.: GlobalInsight: An LSTM based model for multi-vehicle trajectory prediction. In: ICC 2020–2020 IEEE International Conference on Communications (ICC), pp. 1–7. IEEE (2020). https://doi.org/10.1109/ICC40277.2020.9149261
81. Abdellah, A.R., Koucheryavy, A.: VANET traffic prediction using LSTM with deep neural network learning. In: NEW2AN/ruSMART -2020. LNCS, vol. 12525, pp. 281–294. Springer, Cham (2020). https://doi.org/10.1007/978-3-030-65726-0_25
82. Diehl, F., Brunner, T., Le, M.T., Knoll, A.: Graph neural networks for modelling traffic participant interaction. In: 2019 IEEE Intelligent Vehicles Symposium (IV), pp. 695–701. IEEE (2019)
83. Krajewski, R., Bock, J., Kloeker, L., Eckstein, L.: The highd dataset: A drone dataset of naturalistic vehicle trajectories on German highways for validation of highly automated driving systems. In: 2018 21st International Conference on Intelligent Transportation Systems (ITSC), pp. 2118–2125. IEEE (2018). https://doi.org/10.1109/ITSC.2018.8569552

84. Hu, H., Lee, M.J.: Graph neural network-based clustering enhancement in VANET for cooperative driving. In: 2022 International Conference on Artificial Intelligence in Information and Communication (ICAIIC), pp. 162–167. IEEE (2022). https://doi.org/10.1109/ICAIIC54071.2022.9722625

85. Balasubramanian, L., Wurst, J., Botsch, M., Deng, K.: Traffic scenario clustering by iterative optimisation of self-supervised networks using a random forest activation pattern similarity. In: 2021 IEEE Intelligent Vehicles Symposium (IV), pp. 682–689. IEEE (2021)

86. Liang, L., Ye, H., Li, G.Y.: Toward intelligent vehicular networks: A machine learning framework. IEEE Internet Things J. 6(1), 124–135 (2018)

87. Qiao, M., Zhao, H., Zhou, L., Zhu, C., Huang, S.: Topology-transparent scheduling based on reinforcement learning in self-organized wireless networks. IEEE Access 6, 20221–20230 (2018)

88. Zyout, I., Abdel-Qader, I.: Classification of microcalcification clusters via PSO-KNN heuristic parameter selection and GLCM features. Int. J. Comput. Appl. 31(2), 34–39 (2011)

89. Moso, J.C., Cormier, S., Fouchal, H., de Runz, C., Wandeto, J.: Trajectory user linking in C-ITS data analysis. In: GLOBECOM 2020–2020 IEEE Global Communications Conference, pp. 1–6. IEEE (2020). https://doi.org/10.1109/GLOBECOM42002.2020.9322253

90. Argoverse Dataset. https://www.argoverse.org/

A Machine Learning Based Approach to Detect Stealthy Cobalt Strike C&C Activities from Encrypted Network Traffic

Fabian Martin Ramos[1,2] and Xinyuan Wang[1(✉)]

[1] George Mason University, Fairfax, VA 22030, USA
{fmartinr,xwangc}@gmu.edu
[2] ETSI Telecomunicación, Universidad Politécnica de Madrid, 28040 Madrid, Spain

Abstract. Cobalt Strike is a stealthy and powerful command and control (C&C) framework that has been widely used in many recent massive data breach attacks (e.g., the SolarWinds attack in 2020) and ransomware attacks. While detecting Cobalt Strike C&C network traffic is crucial to the protection our mission critical systems from many sophisticated cyberattacks, no existing intrusion detection systems have been shown to be able to reliably detect real world Cobalt Strike C&C traffic from encrypted traffic.

In this paper, we propose a machine learning based approach to detect stealthy Cobalt Strike C&C traffic. Based on the analysis of real world Cobalt Strike traffic, we have developed an approach using flow-level features that capture the inherent characteristics of Cobalt Strike C&C traffic. We have validated our machine learning based detection with five machine learning algorithms and evaluated them with Cobalt Strike traffic from real world cyberattacks. Our empirical results demonstrate that our random forest model can detect close to 50% of real world Cobalt Strike C&C traces in encrypted traffic with a 1.4% false positive rate.

Keywords: Cobalt Strike C&C · Intrusion Detection · Machine Learning

1 Introduction

As our society is increasingly dependent on digital technologies, cyberattacks have become a more serious threat to mission critical systems and infrastructures. For example, recent massive data breach attacks [1] on various business organizations and government agencies (e.g., Target, JP Morgan Chase, OPM, Anthem Inc., Equifax) have impacted tens or hundreds of millions of people. Now cybercriminals can launch sophisticated attacks and compromise mission critical systems via the Internet from virtually anywhere in the world. By using covert command & control (C&C) channels, cybercriminals can stealthily control the compromised systems from thousands of miles away and exfiltrate sensitive data

É. Renault and P. Mühlethaler (Eds.): MLN 2022, LNCS 13767, pp. 113–129, 2023.
https://doi.org/10.1007/978-3-031-36183-8_8

for months. This type of sophisticated cyberattacks are called advanced persistent threats (APT) and they have caused prohibitive financial losses to many businesses in recent times. According to the 2019 Cost of a Data Breach Report [2], the average damage cost of a data breach of 50 million records is $388 million. Specifically, the 2017 Equifax data breach has costed Equifax nearly $1.4 billion as of May 2019 [3]. Cybersecurity Ventures predicted that the annual global cybercrime damage would grow from $3 trillion in 2015 to $10.5 trillion by 2025 [4].

Cobalt Strike has been widely used in recent sophisticated cyberattacks [5,6] due to its ability to establish stealthy C&C channels between the victim system and the attacker. According to Cisco Talos Incident Response (CTIR) Quarterly Report [7], ransomware has been the dominating threat in 2020, and 66% of all ransomware attacks in summer 2020 used Cobalt Strike. The 2020 SolarWinds supply chain attack [8] delivered a customized Cobalt Strike payload to 18,000 Orion customers that included many Fortune 500 organizations (e.g., Microsoft, Cisco) and various US government agencies (e.g., DoD, DHS, DOJ). Cobalt Strike stealthy in-memory persistence and C&C channels enabled the SolarWinds attack to surreptitiously explore, identify and exfiltrate highly sensitive information from some of the most secured information systems without being detected for over nine months. Specifically, the SolarWinds attacker gained access to and exfiltrated the emails of the highest-ranking officials in the Treasure Department [9] and stole gigabytes of source code of Microsoft, Cisco, SolarWinds and FireEye [10].

Detecting Cobalt Strike C&C channel is crucial to the protection of our mission critical cyber systems and infrastructures from many sophisticated cyberattacks. However, Cobalt Strike C&C is very stealthy and hard to detect. Specifically, Cobalt Strike C&C traffic is fully encrypted and can mimic legitimate network traffic using communication protocols such as HTTPS, which will defeat any content based detection. An independent study sponsored by IBM Security [2] shows that it took average 206 days to detect a data breach in 2019. DHS CISA spokesperson Sara Sendek acknowledged that none of deployed intrusion detection or prevention systems (IDS/IPS) – including the U.S. government's multi-billion dollar detection system, Einstein, was able to detect the 2020 SolarWinds breach [11]. In other words, no existing IDS/IPS was able to detect the stealthy Cobalt Strike C&C activities involved in the 2020 SolarWinds attack. To the best of our knowledge, there is no published result on reliable detection of real world stealthy Cobalt Strike C&C activities from encrypted traffic.

In this paper, we explore a new direction in detecting stealthy Cobalt Strike C&C activities from encrypted traffic. Instead of using the packet flow content information, we build our machine learning based detection upon the packet timing and flow duration information that are not changed by encryption. Based on analysis of real world Cobalt Strike C&C traces, we have generated machine learning features that capture the inherent flow level characteristics of encrypted Cobalt Strike C&C traffic. We have empirically validated our machine learning based detection with five popular machine learning models and real world encrypted Cobalt Strike C&C traffic mixed with normal traffic. Our empirical results demonstrate that our our random forest model can detect close to 50% of real world Cobalt Strike C&C traces in encrypted traffic with a 1.4% false positive rate. Our naïve Bayes model achieves over 84% detection true positive rate with 13% false positive rate.

The rest of the paper is structured as follows: Sect. 2 describes and analyze the characteristics of the Cobalt Strike framework and its C&C traffic. Section 3 details the threat model, flow-based features, machine learning models and metrics used in the generation of the model. Section 4 presents the experimental results using real world Cobalt Strike C&C traffic. Section 5 reviews related works. Finally, Sect. 6 concludes the paper with potential future work directions.

2 Analysis of Cobalt Strike C&C

2.1 Cobalt Strike

Cobalt Strike is a powerful and stealthy command and control (C&C) framework that was originally developed by Raphael Smudge as a penetration testing tool in 2012 [12]. Cobalt Strike allows attackers to 1) perform target reconnaissance by identifying known vulnerabilities in software versions; 2) create trojans and malicious website clones that enable drive-by attacks; 3) deploy and inject malicious agents called Beacons into vulnerable targets; 4) perform tasks in the systems infected with a Beacon, such as log keystrokes, take screenshots, execute commands, download additional malware or inject a Beacon in other processes; and 5) disguise its C&C traffic using encryption and mimic normal network traffic using communication protocols such as HTTP, HTTPS or DNS to surpass cybersecurity defenses. Due to its post-exploitation and stealthy C&C features, Cobalt Strike has been widely used in sophisticated cyberattacks.

Cobalt Strike has two main components: the team server and the client. The team server is the C&C server that interacts with a victim that has been infected with a Beacon, and it also accepts client connections. The client is the system used by the attacker to interact with the team server to send commands to the Beacons. Cobalt Strike Beacon is the malware payload used by Cobalt Strike to create a backdoor on a victim system that connects to the team server and can be divided into two parts: the payload stage, and the payload stager. The payload stager is a smaller program that is used to download the payload stage on to a system, inject it into memory, and pass the execution to it. The payload stage is the actual backdoor that runs in memory and can establish a connection to the C&C server through different channels. Cobalt Strike contains many additional components [6], which allow it to be configured to bypass defense systems:

1. Listeners: the listeners define how the Beacon connects to the team server, such as the IP address of the C&C server, the ports and the protocol used. Cobalt Strike supports a great variety of protocols: HTTP, HTTPS and DNS are the most popular ones, while also supporting SMB, raw TCP, forcign listeners (using Metasploit's Meterpeter) and external C&C listeners.

2. Arsenal Kit: these kits allow additional customization into Cobalt Strike capabilities to evade antivirus products. Some of the most popular kits are:
   - Artifact kit: allows attackers to modify the template for all Cobalt Strike executables, DLLs and shellcode.
   - Elevate kit: allows attackers to include third-party privilege escalation scripts with Cobalt Strike Beacon.
   - Mimikatz kit: allows attackers to use and update the Mimikatz installation included with Cobalt Strike.
3. Malleable C&C profiles: part of the Arsenal Kit, it allows the attacker to customize the communications between the Beacon and the team server as well as the Beacon in-memory characteristics, determine how it does process injection, and influence post-exploitation jobs.

The Malleable C&C profile's ability to customize the network traffic that the Beacon generates and receives, such as the interval between each Beacon callback to the team server, the URI of the HTTP/S requests, inserting additional data to masquerade the actual data payload size and more, is the main characteristic that makes Cobalt Strike such a powerful tool for penetration testers and malicious agents alike. It allows the attacker to blend the C&C traffic with the normal traffic, bypassing security defenses such as network firewalls and intrusion detection systems.

Cobalt Strike has become one of the favorite tools for attackers of all skill levels, from script kiddies to state-sponsored attackers, being used among other malware such as QBot or Emotet and phishing attacks or being actively used for all the phases of the attack lifecycle. Cobalt Strike has also been identified not only in ransomware attacks, but also part of the famous SolarWinds supply chain attack in 2020 and cyberespionage campaigns targeting the Ukrainian population.

2.2 Cobalt Strike C&C Communications

This paper focuses on the detection of the two most popular application level protocols used by attackers for Cobalt Strike C&C communications: HTTP and HTTPS. Per the Cobalt Strike documentation, a typical Cobalt Strike C&C HTTP/S transaction between an infected host and the C&C server proceeds as follows:

- First, the TCP connection is established via the TCP three-way handshake (SYN, SYN/ACK, ACK).
- If the HTTPS protocol is used, the TLS handshake is performed (Client Hello, Server Hello, Server Key Exchange, Client Key Exchange, Finished) to establish the HTTPS session.
- Then, the exchange of data takes place between the victim and the team server.
- Finally, the HTTPS session and TCP connection are closed.

The Beacon performs callbacks to the C&C server periodically to retrieve tasks to execute, sending the information of the infected system in a GET request after the connection has been established. The server responds to the request with the tasks, if any have been instructed by the attacker. The Beacon then executes the tasks and initiates another connection to send the output back to the C&C server using a POST request. The server then responds with data that is discarded by the Beacon. The packet payload is encoded and encrypted to complicate the analysis of its contents by security systems. Cobalt Strike uses this process to simulate a legitimate HTTP/S exchange between a client and a server.

Figure 1 shows several HTTPS transactions between a Beacon-infected host and a team server from a packet capture using the network protocol analyzer tool Wireshark. In the image, it is possible to observe that, after establishing the HTTPS session through the TLS Handshake, the victim and the C&C server exchange "Application Data" packets which are used to transmit the tasks that are to be executed to the Beacon or send the results from those tasks back to the C&C server. Once the data has been exchanged, the C&C server closes the HTTPS session using an "Encrypted Alert" packet. Each connection between the victim and the team server requires the client to open a new HTTPS session with the server.

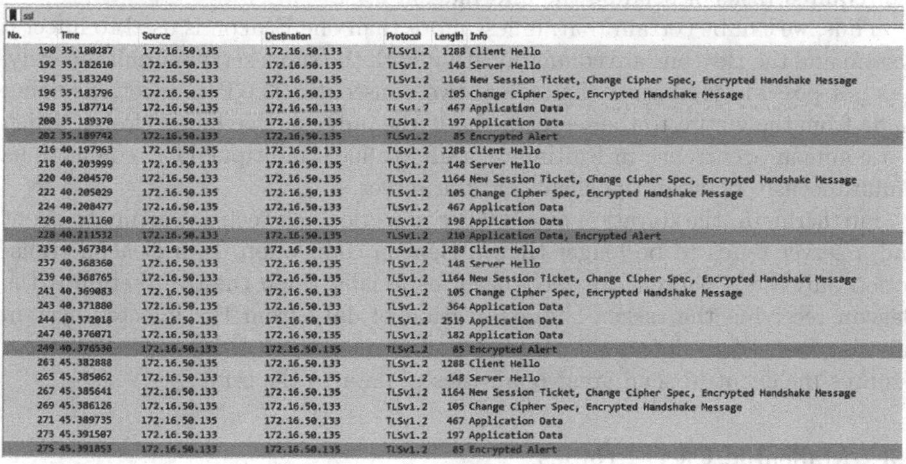
ssl

No.	Time	Source	Destination	Protocol	Length	Info
190	35.180287	172.16.50.135	172.16.50.133	TLSv1.2	1288	Client Hello
192	35.182610	172.16.50.133	172.16.50.135	TLSv1.2	148	Server Hello
194	35.183249	172.16.50.133	172.16.50.135	TLSv1.2	1164	New Session Ticket, Change Cipher Spec, Encrypted Handshake Message
196	35.183796	172.16.50.135	172.16.50.133	TLSv1.2	105	Change Cipher Spec, Encrypted Handshake Message
198	35.187714	172.16.50.135	172.16.50.133	TLSv1.2	467	Application Data
200	35.189370	172.16.50.133	172.16.50.135	TLSv1.2	197	Application Data
202	35.189742	172.16.50.133	172.16.50.135	TLSv1.2	85	Encrypted Alert
216	40.197963	172.16.50.135	172.16.50.133	TLSv1.2	1288	Client Hello
218	40.203999	172.16.50.133	172.16.50.135	TLSv1.2	148	Server Hello
220	40.204570	172.16.50.133	172.16.50.135	TLSv1.2	1164	New Session Ticket, Change Cipher Spec, Encrypted Handshake Message
222	40.205029	172.16.50.135	172.16.50.133	TLSv1.2	105	Change Cipher Spec, Encrypted Handshake Message
224	40.208477	172.16.50.135	172.16.50.133	TLSv1.2	467	Application Data
226	40.211160	172.16.50.133	172.16.50.135	TLSv1.2	198	Application Data
228	40.211532	172.16.50.133	172.16.50.135	TLSv1.2	210	Application Data, Encrypted Alert
235	40.367384	172.16.50.135	172.16.50.133	TLSv1.2	1288	Client Hello
237	40.368360	172.16.50.133	172.16.50.135	TLSv1.2	148	Server Hello
239	40.368765	172.16.50.133	172.16.50.135	TLSv1.2	1164	New Session Ticket, Change Cipher Spec, Encrypted Handshake Message
241	40.369083	172.16.50.135	172.16.50.133	TLSv1.2	105	Change Cipher Spec, Encrypted Handshake Message
243	40.371880	172.16.50.135	172.16.50.133	TLSv1.2	364	Application Data
244	40.372018	172.16.50.135	172.16.50.133	TLSv1.2	2519	Application Data
247	40.376071	172.16.50.133	172.16.50.135	TLSv1.2	182	Application Data
249	40.376530	172.16.50.133	172.16.50.135	TLSv1.2	85	Encrypted Alert
263	45.382888	172.16.50.135	172.16.50.133	TLSv1.2	1288	Client Hello
265	45.385062	172.16.50.133	172.16.50.135	TLSv1.2	148	Server Hello
267	45.385641	172.16.50.133	172.16.50.135	TLSv1.2	1164	New Session Ticket, Change Cipher Spec, Encrypted Handshake Message
269	45.386126	172.16.50.135	172.16.50.133	TLSv1.2	105	Change Cipher Spec, Encrypted Handshake Message
271	45.389735	172.16.50.135	172.16.50.133	TLSv1.2	467	Application Data
273	45.391507	172.16.50.133	172.16.50.135	TLSv1.2	197	Application Data
275	45.391853	172.16.50.133	172.16.50.135	TLSv1.2	85	Encrypted Alert

Fig. 1. Packet capture from HTTPS Beacon transaction

Comparing the previous figure with Fig. 2, which shows a packet capture of legitimate HTTPS traffic when accessing a web pages such as Facebook (www.facebook.com), we can observe some notable differences between legitimate traffic and Beacon traffic.

First, the Beacon does not use the same TCP connection to contact the C&C server more than once. This event occurs for both the HTTP and HTTPS

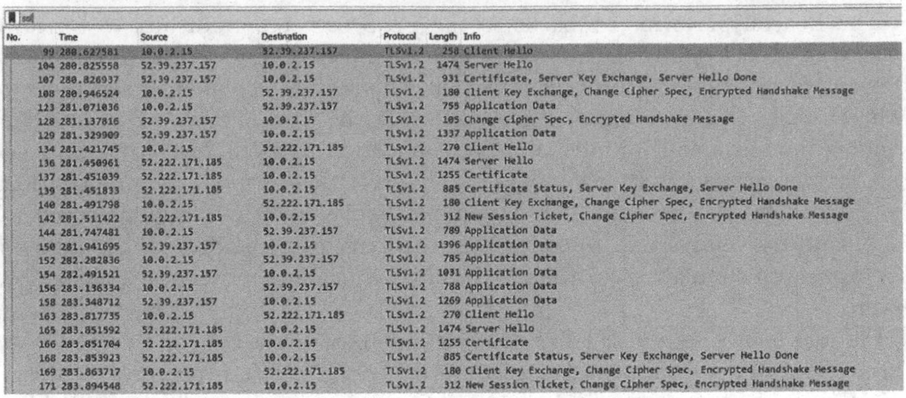

No.	Time	Source	Destination	Protocol	Length	Info
99	280.627581	10.0.2.15	52.39.237.157	TLSv1.2	258	Client Hello
104	280.825558	52.39.237.157	10.0.2.15	TLSv1.2	1474	Server Hello
107	280.826937	52.39.237.157	10.0.2.15	TLSv1.2	931	Certificate, Server Key Exchange, Server Hello Done
108	280.946524	10.0.2.15	52.39.237.157	TLSv1.2	180	Client Key Exchange, Change Cipher Spec, Encrypted Handshake Message
123	281.071036	10.0.2.15	52.39.237.157	TLSv1.2	755	Application Data
128	281.137816	52.39.237.157	10.0.2.15	TLSv1.2	105	Change Cipher Spec, Encrypted Handshake Message
129	281.329909	52.39.237.157	10.0.2.15	TLSv1.2	1337	Application Data
134	281.421745	10.0.2.15	52.222.171.185	TLSv1.2	270	Client Hello
136	281.450961	52.222.171.185	10.0.2.15	TLSv1.2	1474	Server Hello
137	281.451039	52.222.171.185	10.0.2.15	TLSv1.2	1255	Certificate
139	281.451833	52.222.171.185	10.0.2.15	TLSv1.2	885	Certificate Status, Server Key Exchange, Server Hello Done
140	281.491798	10.0.2.15	52.222.171.185	TLSv1.2	180	Client Key Exchange, Change Cipher Spec, Encrypted Handshake Message
142	281.511422	52.222.171.185	10.0.2.15	TLSv1.2	312	New Session Ticket, Change Cipher Spec, Encrypted Handshake Message
144	281.747481	10.0.2.15	52.39.237.157	TLSv1.2	789	Application Data
150	281.941695	52.39.237.157	10.0.2.15	TLSv1.2	1396	Application Data
152	282.282836	10.0.2.15	52.39.237.157	TLSv1.2	785	Application Data
154	282.491521	52.39.237.157	10.0.2.15	TLSv1.2	1031	Application Data
156	283.136334	10.0.2.15	52.39.237.157	TLSv1.2	788	Application Data
158	283.348712	52.39.237.157	10.0.2.15	TLSv1.2	1269	Application Data
163	283.817735	10.0.2.15	52.222.171.185	TLSv1.2	270	Client Hello
165	283.851592	52.222.171.185	10.0.2.15	TLSv1.2	1474	Server Hello
166	283.851704	52.222.171.185	10.0.2.15	TLSv1.2	1255	Certificate
168	283.853923	52.222.171.185	10.0.2.15	TLSv1.2	885	Certificate Status, Server Key Exchange, Server Hello Done
169	283.863717	10.0.2.15	52.222.171.185	TLSv1.2	180	Client Key Exchange, Change Cipher Spec, Encrypted Handshake Message
171	283.894548	52.222.171.185	10.0.2.15	TLSv1.2	312	New Session Ticket, Change Cipher Spec, Encrypted Handshake Message

Fig. 2. Captured HTTPS Packets from Normal Traffic

Beacons, but it can be more clearly observed in the case of the HTTPS Beacon: the C&C server sends an "Encrypted Alert" packet indicating that the HTTPS session is closed, and consequently also closes the TCP connection shortly after that packet is sent. Therefore, as we can also observe in Fig. 1, the Beacon opens a TCP connection for every call to the C&C server and closes it after retrieving the required tasks or sending the information back.

Thus, we can be certain that, unless more than one Beacon is used to infect a system and the Beacons are communicating with the same server simultaneously, it is not possible to observe the case of two consecutive TCP connections being made from the victim to the server without the previous one being closed, which is a common occurrence in legitimate traffic to have multiple TCP connections simultaneously between the client and the server.

Furthermore, the duration of a TCP connection between a legitimate client and a server tends to be longer than a Beacon connection. In a Beacon transaction only a few packets are exchanged, especially when the transaction is the Beacon receiving the tasks. The exfiltration of data from the infected system can also be used to detect the Beacon traffic, since the legitimate traffic rarely requires the client to send great quantities of data to the server.

2.3   Malleable C&C Profiles

Cobalt Strike introduces the use of Malleable C&C profiles to customize the network indicators of the C&C traffic between the Beacon and the team server as an evasion measure, as they can be used to disguise the Beacon traffic to look like other malware or blend-in with legitimate traffic, making it difficult to detect. The Malleable C&C profiles are loaded onto the team server and modify the in-memory characteristics of the Beacon, how to transform and store the data in a transaction and post-exploitation functions.

The Malleable C&C profile is structured into several sections which are used to configure the global Beacon behavior or specific behavior depending on the communication protocol used. The global configuration includes the *sleeptime* which is the Beacon callback interval, the *data jitter* which is a random length string up to the chosen value (in bytes) that is appended to the server value and the *useragent* which sets the User-Agent string used in the HTTP requests identifying the application, operating system, vendor and version of the requesting computer program.

The Malleable C&C profile can also be used to configure the HTTP headers, the SSL certificate used in the HTTPS sessions, the URI of the HTTP requests, the HTTP verb used in the transactions and the modifications performed to the payload of the packets sent from the server and the Beacon, including appended data and encoding.

Our analysis of the HTTP/S Cobalt Strike traffic and the Malleable C&C profiles has identified several network indicators that can be used to detect the Cobalt Strike C&C traffic: 1) the Beacon sleeps for an interval of time after not receiving tasks or after sending the output to the server; 2) most of the sessions have a short duration and few data packets are exchanged; 3) in most cases the victim sends more data to the C&C server than the opposite, especially when it is exfiltrating information about the victim.

3   A Machine Learning Based Detection

3.1   Threat Model

Our objective is to build a machine learning based detection system that can identify and detect stealthy Cobalt Strike C&C activity from encrypted network traffic in near real-time. We assume the following:

- The system has already been infected with a Cobalt Strike Beacon and that the initial infection has not been detected by IDS/IPS.
- The Cobalt Strike Beacon initiates the communications with the C&C server after completing the download of the Beacon payload.
- The attackers use encryption and customized Cobalt Strike Malleable C&C profiles to mimic legitimate traffic and evade detection. Consequently, the C&C packets exchanged do not have any fixed pattern.

3.2   Flow Based Features

In order to detect Cobalt Strike C&C traffic amongst legitimate traffic, we have used the traffic analysis software Zeek [13] to extract the network indicators of the individual flows or connections. Zeek produces a record for each connection that has occurred with a system in the log file *conn.log* in real-time, but it can also be used to analyze packet captures and output the connections within. While a connection is usually associated with the TCP protocol, Zeek can also track stateless protocols like UDP. The flow-related features that have been selected from the Zeek output are the following:

- **id.orig_h:** string. IP address of the host that initiated the connection.
- **id.orig_p:** integer. Port used by the host that initiated the connection.
- **id.resp.h:** string. IP address of the host that received the connection.
- **id.resp.p:** integer. Port used by the host that received the connection.
- **proto:** string. Transport layer protocol of the connection (TCP, UDP, ICMP).
- **service:** string. Identification of an application protocol sent over the connection (DNS, HTTP, HTTPS).
- **duration:** double. Total duration of the connection.
- **orig_bytes:** integer. Payload bytes that the originator of the connection sent.
- **resp_bytes:** integer. Payload bytes that the responder sent.
- **conn_state:** string. State of the connection, some typical values include SHR (responder sent SYN ACK followed by SYN) and S0 (connection attempt seen).
- **history:** string. Records the state history of connections as a string of letters.
- **orig_pkts:** integer. Number of packets that the originator sent during the connection.
- **orig_ip_bytes:** integer. Number of IP level bytes that the originator sent during the connection.
- **resp_pkts:** integer. Number of packets that the responder sent during the connection.
- **resp_ip_bytes:** integer. Number of IP level bytes that the responder sent during the connection.

Features such as the IP addresses and ports of the sender and the receiver of the connection will be used to identify the connections, but cannot be used in the detection process. Other features such as the protocol, duration of the connection, packets and bytes exchanged will be used by the machine learning model to make the predictions on the traffic, based on the analysis of the Cobalt Strike C&C traffic.

3.3   Machine Learning Models

In order to select the machine learning algorithms that will be used to develop the model, it will be necessary to assess which algorithms best tackle the problem at hand. Since the goal of the model is to make predictions on the network traffic to determine if a system has been infected with a Cobalt Strike Beacon, we can identify it as a binary classification problem. However, since a system being infected with Cobalt Strike is considered an abnormal event due to its rare occurrence probability, it can also be identified as an anomaly detection problem, even if the Beacon traffic features do not differ significantly from the legitimate traffic, as its purpose it to disguise itself as such. The training dataset records will be labeled, thus both supervised and unsupervised anomaly detection analysis can be performed. However, as we will observe, the unsupervised model will obtain worse results due to the similarity in feature values that the Beacon and the legitimate traffic have.

The following machine learning algorithms will be evaluated to generate the model: random forest, artificial neural network, support vector machine and naïve Bayes for the supervised model, and K-Means clustering algorithm for the unsupervised model.

3.4   Evaluation Metrics

A variety of metrics will be used to objectively evaluate the different machine learning algorithms that will be used to generate the model that will make predictions on network traffic to discover if a system has been infected with a Cobalt Strike Beacon.

A confusion matrix is a table commonly used to describe the performance of a machine learning model when performing classification tasks over a test dataset whose labels (or classes) are known. It does so by establishing a relationship between the real label of a record and the predicted label of the same record, thus graphically exposing the number of records that have been correctly and incorrectly classified.

Independently of the number of labels in the dataset, the confusion matrix outputs four values for each label. For a specific label, the number of true positives (TP) is the number of records corresponding to that label that have been correctly classified as such. The number of false negatives (FN) is the number of records corresponding to the label that have been misclassified as belonging to other labels. The number of false positives (FP) indicates the number of records belonging to different labels that have been misclassified as belonging to the chosen label. Last, the number of true negatives (TN) is the number of records that have been correctly classified as belonging to other labels. In the case of our paper, the positive label will refer to the Beacon traffic, while the legitimate traffic will be labeled as negative.

Using the output of the confusion matrix, it is possible to calculate the following metrics to evaluate the machine learning algorithms:

- **Accuracy:** Ratio between the number of correct predictions and the total number of predictions made. It is the best indicator of the algorithm's performance only if the dataset has the same number of records for each class.

$$Accuracy = \frac{TP + TN}{TP + FP + TN + FN} \tag{1}$$

- **Precision:** The precision is the ratio between the number of correct predictions for a class and the total number of predictions of said class. The weighted precision is used to evaluate the model for all classes, calculating a weighted average depending on the probability of occurrence of each class.

$$Precision = \frac{TP}{TP + FP} \tag{2}$$

- **Recall or True Positive Rate (TPR):** Proportion of positive records correctly classified as such compared to the total number of positive records.

It can be considered a percentage. Similarly to the precision, we will consider the weighted recall for all the labels.

$$TPR = \frac{TP}{TP + FN} \tag{3}$$

- **False Positive Rate (FPR):** Proportion of negative records incorrectly classified as positive records, with respect to all negative records. It can be considered a percentage.

$$FPR = \frac{FP}{FP + TN} \tag{4}$$

- **F1 score:** Measures the performance of a model by considering both its precision as well as its robustness. To do so, it uses the Precision and Recall values.

$$F1 = 2 \cdot \frac{Precision \cdot Recall}{Precision + Recall} \tag{5}$$

When evaluating the model certain evaluation metrics will have more importance than others. That will be the case of the true positive rate or TPR, which will indicate the model's ability to correctly detect the Beacon traffic. The false positive rate or FPR will be used to evaluate the adequacy of the model to be deployed in a real environment, as most of the network traffic that an intrusion detection system analyzes is legitimate, a low rate of false alarms is required for its deployment. For example, if a detection system is deployed in a real environment that will potentially see hundreds of thousands to millions of connections, a high false positive rate will cause disruptions in the legitimate activity of the users of the network, as tens of thousands of legitimate connections will be detected as malicious. Finally, the F1 score will provide overall information about the model's accuracy and robustness.

4   Empirical Validation

4.1   Dataset Acquisition

The machine learning based detection requires a training dataset to develop the machine learning model and a testing dataset to evaluate the performance of the model. Both datasets contain legitimate traffic and Cobalt Strike C&C traffic, so that the model is developed and evaluated using traffic that is closely related with the traffic that would be found in a real life environment.

The legitimate traffic has been obtained from the popular cybersecurity dataset CICIDS17 [14] of the Canadian Institute of Cybersecurity and the public CTU-Normal-20 dataset [15] of the Stratosphere Research Laboratory. For the training dataset, we have generated and collected 31 Cobalt Strike C&C packet captures (PCAP) in our lab environment using different Malleable C&C profiles that emulate advanced penetration threats (APT), crimeware, normal traffic, or generated using randomizers such as [16]. The use of different Malleable C&C

profiles and commands reduces the possibility of overfitting the model to a specific profile, thus enabling the machine learning models to perform better in the presence of previously unseen real world Cobalt Strike traffic. The testing dataset, on the other hand, has been generated using 25 captured traces of real world cyberattacks that include Cobalt Strike C&C traffic from malware-traffic-analysis.net [17]. All the Cobalt Strike traffic in these traces has been manually labeled by cybersecurity teams, which enables us to measure the performance of our machine learning based detection.

Each record and its features in the dataset corresponds to an individual flow or connection in the network, which has been extracted from the acquired traffic using the network analysis tool Zeek. Each record has been labeled using an additional feature, *label*, which will take two values: 1 if the record corresponds to Cobalt Strike Beacon traffic, and 0 if the record corresponds to other types of traffic (legitimate or unknown, this is due to some of the test captures containing malicious traffic generated from other malware).

Table 1. Dataset Content

	Cobalt Strike (HTTP)	Cobalt Strike (HTTPS)	Legitimate HTTPS
Training dataset	5,500 records	4,000 records	391,500 records
Testing dataset	450 records	3,150 records	10,800 records

Table 1 shows the resulting record distribution of the datasets for each of the labels. Given that Cobalt Strike C&C traffic is very rare in real world, we deliberately build the training dataset with significantly more legitimate traffic records than Beacon traffic records. Such a training dataset not only enables the machine learning model to detect Cobalt Strike activities in real world scenarios, but also helps reduce potential detection false positives generated by the model. The percentage of Beacon records, which is around 2% of all the records in the training dataset, is a lower than average percentage for most cybersecurity datasets, but it is enough for most models to be able to detect the Beacon traffic generated using different Malleable C&C profiles and also correctly identify the legitimate traffic producing a low rate of false alarms.

4.2 Flow Based Machine Learning Detection

First, we selected the features from the raw dataset that will be used in the detection of the network traffic. Each record belonging to the dataset has 14 features out of which four of those features (*src_ip*, *src_p*, *dst_ip* and *dst_p*) will only be used for identification purposes. The IP addresses and ports used in the connection communications will not be used to detect the Beacon traffic, as they are network indicators that can easily be changed by the attacker, thus evading the detection system.

Since the machine learning model will focus on the detection of HTTP/S Beacons, which use the TCP protocol in the transport layer, the *proto* and *service* features will be key in the identification of the Beacon traffic. The state of the connection feature, *conn_state*, will not be selected because it will not help in the identification of the Beacon traffic, since it only identifies if the connection has correctly finished or not and most if not all the records in the dataset have the same value for the connection. On the contrary, the duration of the connection will not be used in the detection of the Beacon traffic because it can vary due to external factors such as network latency and packet drops regardless of the type of traffic.

The *history* feature, extracted by Zeek, records the "history" of the transactions in the connection such as the type of packets that were transmitted, the order in which the packets were sent, and who send the packets by using letters. Therefore, based on the analysis of the Beacon traffic performed in previous sections, we can assume that the history feature will have similar values for the connections made by the Beacon traffic, as it always follows the same general schema of communications. The selection of the rest of the features (*orig_bytes*, *resp_bytes*, *orig_pkts*, *orig_ip_bytes*, *resp_pkts*, *resp_ip_bytes*) will be done alongside the evaluation of the model, depending on how they affect the performance of the model.

Since several features selected contain categorical data - *proto*, *service* and *history* more precisely - it will be necessary to perform feature encoding to transform their values to numerical data. To do so, we transformed the features' values from string to integer by assigning an integer value to each distinct string value according to the frequency of appearance of the value and generating an additional feature with "_index" in the name. If a string value appears for the first time in the testing dataset, it will be assigned the same value as the least frequent value.

The hyperparameter tuning phase has been performed using the results from the machine learning model evaluation for their respective algorithms. However, all the tested values for the tuning phase will not be explained, since most experiments regarding the hyperparameter tuning phase give the reader little information, as they consist in manual changes to the hyperparameter values which improve the performance of the models by less than 1% on most occasions.

The random forest model has been generated using *proto_index*, *service_index*, *history_index*, *orig_bytes*, *resp_bytes*, *orig_pkts*, *orig_ip_bytes* and *resp_pkts* as input features, and the hyperparameters values of a maximum depth of 15 and 30 trees created. As we can observe, the random forest model achieves a moderately good detection rate (around 50%) and a low false alarm rate of 1.4%. On the other hand, the neural network model achieves poor detection results with a 3% detection rate but a low false alarm rate of 0.3%. The neural network model has been built using a four-layer structure, using two hidden layers with 18 nodes each, taking the 9 selected features as inputs (*proto_index*, *service_index*, *history_index*, *orig_bytes*, *resp_bytes*, *orig_pkts*, *orig_ip_bytes*, *resp_pkts* and *resp_ip_bytes*) and utilizes the L-BFGS solver.

The naïve Bayes model and the linear support vector machine achieve higher detection rate than the random forest, at the expense of the false positive

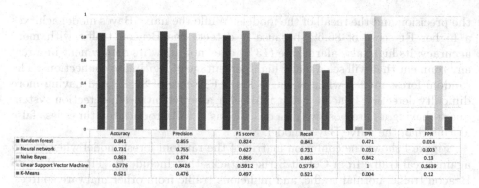

	Accuracy	Precision	F1 score	Recall	TPR	FPR
■ Random forest	0.841	0.855	0.824	0.841	0.471	0.014
■ Neural network	0.731	0.755	0.627	0.731	0.031	0.003
■ Naïve Bayes	0.863	0.874	0.866	0.863	0.842	0.13
■ Linear Support Vector Machine	0.5776	0.8426	0.5912	0.5776	1	0.5639
■ K-Means	0.521	0.476	0.497	0.521	0.004	0.12

Fig. 3. Comparison of the performance of the machine learning models

rate. The naïve Bayes model takes *proto_index*, *service_index*, *history_index*, *orig_bytes* and *orig_pkts* as input features and using a multinomial model. While obtaining a high true positive rate at 84%, it does not serve a real intrusion detection system due to having a 13% false positive rate. Similarly, the linear support vector uses three input features *proto_index*, *service_index*, *history_index* and an aggregation depth of 4 achieving a 100% detection rate but misclassifying 56% of the legitimate traffic. Finally, the unsupervised clustering machine learning model K-means attempts to group the Beacon and legitimate traffic in two separate clusters using the same input features as the linear support vector machine model, but ultimately fails to do so, obtaining a 0.4% true positive rate and a 12% false negative rate.

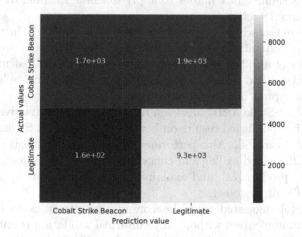

Fig. 4. Confusion Matrix of the Random Forest Model

As shown in Fig. 3, the random forest and naïve Bayes models perform better than the rest of the models if we consider the F1 score, which considers both the precision and the recall of the models. While the naïve Bayes model achieves a higher F1 score of 86.6% because it detects the Beacon traffic with more accuracy, its high false alarm rate (13%) does not allow its deployment in a real environment that will see a really high percentage of legitimate connections. The random forest model, while having a lower F1 score of 82.4% and having more difficulty detecting Beacon traffic, makes for a better intrusion detection system due to having a false positive rate of 1.4%, as it will produce ten times less false alarms than the naïve Bayes model.

Figure 4 shows the confusion matrix of the random forest model when tested against real traffic from Cobalt Strike attacks that include HTTP and HTTPS Beacon traffic, normal traffic, and malicious traffic from other malware sources. Since the focus of this work is to detect Cobalt Strike C&C traffic, the malicious traffic from other sources has been labeled as legitimate.

5   Related Works

Machine learning techniques (e.g., decision trees, neural network) have a long history of being used in constructing Intrusion Detection Systems (IDS) [18–23]. Recent machine learning based IDS approaches have been shown to have good accuracy and acceptable efficiency in detecting/classifying attacks in the public datasets (e.g., UNSW-NB15, CICIDS17). Yet, machine learning based approaches have achieved far less success in real world intrusion detection than in other areas such as speech recognition. R. Sommer et al. [24] examined the fundamental differences between the network intrusion detection problem and those problems where machine learning regularly finds much more success, and argued that it is significantly harder to apply machine learning techniques effectively in intrusion detection.

Most proposed machine learning based IDS approaches have focused on detecting known exploits rather than stealthy C&C activities from encrypted traffic. Gardiner et al. [25] examined evasion techniques against machine learning based detection, and pointed out that many existing machine learning based C&C detection approaches are vulnerable to evasion.

Despite Cobalt Strike C&C has been used in almost all massive breaches [5], there are very few published results on how to detect Cobalt Strike C&C traffic. Navarrete et al. from Palo Alto performed an extensive analysis on the Cobalt Strike C&C traffic encoding [26] and encryption [27] of the payload as well as the Malleable C&C profiles [28], and explained why such versatility makes Cobalt Strike C&C difficult to detect.

B. Vennyk [5] suggested using beaconing characteristics to detect Cobalt Strike C&C communication without any empirical validation result. In addition, it did not consider the data jitter and the sleep jitter that the attackers can configure to significantly change the beaconing pattern.

N. Kanzig et al. [29] proposed a machine learning based approach to identify C&C channels using features extracted by CICFlowMeter. While they have shown that their random forest classifier can detect their lab generated Cobalt

Strike C&C traffic, their work does not demonstrate if their classifier can detect any real world Cobalt Strike C&C traffic captured from real world attacks. In addition, they have not considered the data jitter that can be introduced into the server responses which would impact the packet flow features.

Van der Eijk et al. [30] proposed a threshold based approach detect Cobalt Strike C&C traffic. The proposed method was able to detect the Cobalt Strike C&C traffic generated with a few Malleable profiles in their lab environment.

In summary, all previous machine learning based Cobalt Strike C&C detection approaches used lab generated Cobalt Strike traffic with very few Malleable profiles, and none of them have been validated with real world Cobalt Strike C&C traffic. In contrast, our machine learning based detection has been trained with 31 different Malleable profiles of Cobalt Strike C&C and validated with Cobalt Strike C&C traffic captured from real world cyberattacks.

6   Conclusions

Given the wide spread use of Cobalt Strike C&C in recent massive data breach attacks and ransomware attacks, it is critically important to be able to detect the stealthy Cobalt Strike C&C traffic in order to effectively mitigate the these stealthy and damaging cyberattacks.

In this paper, we propose using flow based features to detect Cobalt Strike Beacon C&C traffic, and evaluate five machine learning algorithms to develop a model that can detect the Beacon traffic. To the best of our knowledge, our machine learning based detection is the first to be validated using Cobalt Strike C&C traffic captured from real world cyberattacks. Our experimental results show that it is feasible to detect real world, previously unseen Cobalt Strike C&C traffic with a reasonable true positive rate and low false alarm rate at the same time.

For future work, we plan to look for more effective machine features for detecting stealthy Cobalt Strike C&C traffic, and investigate how to improve the detection true positive rate and reduce the false positive rate at the same time by combining different machine learning models.

References

1. Swinhoe, D.: The 15 biggest data breaches of the 21st century (2020). https:// www.csoonline.com/article/2130877/the-biggest-data-breaches-of-the-21st-centur y.html
2. Security, I.: Cost of a Data Breach Report 2019 (2019). https://f.hubspotusercon tent40.net/hubfs/2783949/2019_Cost_of_a_Data_Breach_Report_final.pdf
3. Schwartz, M.J.: Equifax's Data Breach Costs Hit $1.4 Billion (2019). https://www. bankinfosecurity.com/equifaxs-data-breach-costs-hit-14-billion-a-12473
4. Morgan, S.: Cybercrime To Cost The World $10.5 Trillion Annually By 2025 (2020). https://cybersecurityventures.com/cybercrime-damages-6-trillion-by-2021/

5. Vennyk, B.: How to Detect CobaltStrike Command & Control Communication (2021). https://underdefense.com/guides/how-to-detect-cobaltstrike-command-control-communication/
6. Rahman, A.: Defining Cobalt Strike Components So You Can BEA-CONfident in Your Analysis (2021). https://www.mandiant.com/resources/blog/defining-cobalt-strike-components
7. Liebenberg, D., Huey., C.: Quarterly Report: Incident Response Trends in Summer 2020 (2020). https://blog.talosintelligence.com/2020/09/CTIR-quarterly-trends-Q4-2020.html
8. FireEye: Highly Evasive Attacker Leverages SolarWinds Supply Chain to Compromise Multiple Global Victims With SUNBURST Backdoor (2020). https://www.fireeye.com/blog/threat-research/2020/12/evasive-attacker-leverages-solarwinds-supply-chain-compromises-with-sunburst-backdoor.html
9. United States federal government data breach. https://en.wikipedia.org/wiki/2020_United_States_federal_government_data_breach
10. Abrams, L.: SolarLeaks site claims to sell data stolen in SolarWinds attacks (2021). https://www.bleepingcomputer.com/news/security/solarleaks-site-claims-to-sell-data-stolen-in-solarwinds-attacks/
11. Timberg, C., Nakashima, E.: The U.S. government spent billions on a system for detecting hacks. The Russians outsmarted it (2021). https://www.seattletimes.com/nation-world/the-u-s-government-spent-billions-on-a-system-for-detecting-hacks-the-russians-outsmarted-it/
12. Software for Adversary Simulations and Red Team Operations. https://www.cobaltstrike.com
13. An Open Source Network Security Monitoring Tool. https://zeek.org/
14. For Cybersecurity, C.I.: Intrusion Detection Evaluation Dataset (CIC-IDS 2017). https://www.unb.ca/cic/datasets/ids-2017.html
15. Laboratory, S.R.: Malware Capture Facility Project. https://www.stratosphereips.org/datasets-normal
16. Malleable-C2-Randomizer. https://github.com/bluscreenofjeff/Malleable-C2-Randomizer
17. A source for packet capture (pcap) files and malware samples. https://www.malware-traffic-analysis.net/index.html
18. Sinclair, C., Pierce, L., Matzner, S.: An application of machine learning to network intrusion detection. In: Proceedings of the 5th Annual Computer Security Applications Conference (ACSAC 1999) (1999)
19. Sangkatsanee, P., Wattanapongsakorn, N., Charnsripinyo, C.: Practical real-time intrusion detection using machine learning approaches. Comput. Commun. **34**(18), 2227–2235 (2011)
20. Kulariya, M., Saraf, P., Ranjan, R., Gupta, G.P.: Performance analysis of network intrusion detection schemes using apache spark. In: Proceedings of the 2016 International Conference on Communication and Signal Processing (ICCSP 2016), pp. 1973–1977. IEEE (2016)
21. Toupas, P., Chamou, D., Giannoutakis, K.M., Drosou, A., Tzovaras, D.: An intrusion detection system for multi-class classification based on deep neural networks. In: Proceedings of the 18th IEEE International Conference On Machine Learning And Applications (ICMLA 2019). IEEE (2019)

22. Quinan, P.G., Traore, I., Gondhi, U.R., Woungang, I.: Unsupervised anomaly detection using a new knowledge graph model for network activity and events. In: Renault, É., Boumerdassi, S., Mühlethaler, P. (eds.) MLN 2021. LNCS, vol. 13175, pp. 117–130. Springer, Cham (2022). https://doi.org/10.1007/978-3-030-98978-1_8

23. Tufan, E., Tezcan, C., Acartürk, C.: Anomaly-based intrusion detection by machine learning: a case study on probing attacks to an institutional network. IEEE Access **9**, 50078–50092 (2021)

24. Sommer, R., Paxson, V.: Outside the closed world: on using machine learning for network intrusion detection. In: Proceedings of the 2010 IEEE Symposium on Security and Privacy (S&P 2010). IEEE (2010)

25. Gardiner, J., Nagaraja, S.: On the security of machine learning in malware C&C detection: a survey. ACM Comput. Surv. **49**(3), 1–39 (2017)

26. Navarrete, C., Sangvikar, D., Guan, A., Fu, Y., Jia, Y., Shibiraj, S.: Cobalt Strike Analysis and Tutorial: CS Metadata Encoding and Decoding (2022). https://unit42.paloaltonetworks.com/cobalt-strike-metadata-encoding-decoding/

27. Navarrete, C., Sangvikar, D., Guan, A., Fu, Y., Jia, Y., Shibiraj, S.: Cobalt Strike Analysis and Tutorial: CS Metadata Encryption and Decryption (2022). https://unit42.paloaltonetworks.com/cobalt-strike-metadata-encryption-decryption/

28. Navarrete, C., Sangvikar, D., Guan, A., Fu, Y., Jia, Y., Shibiraj, S.: Cobalt Strike Analysis and Tutorial: How Malleable C2 Profiles Make Cobalt Strike Difficult to Detect (2022). https://unit42.paloaltonetworks.com/cobalt-strike-malleable-c2-profile/

29. Känzig, N., Meier, R., Gambazzi, L., Lenders, V., Vanbever, L.: Machine learning-based detection of C&C channels with a focus on the locked shields cyber defense exercise. In: Proceedings of the 11th International Conference on Cyber Conflict (CyCon 2019), pp. 1–19. IEEE (2019)

30. van der Eijk, V., Schuijt, C.: Detecting Cobalt Strike Beacons in NetFlow Data. https://rp.os3.nl/2019-2020/p29/report.pdf

Unified Emulation-Simulation Training Environment for Autonomous Cyber Agents

Li Li[1(✉)], Jean-Pierre S. El Rami[1], Adrian Taylor[1], James Hailing Rao[2], and Thomas Kunz[3]

[1] Defence Research and Development Canada, Ottawa, Canada
li.li2@ecn.forces.gc.ca
[2] Queen's University, Kingston, Canada
[3] Carleton University, Ottawa, Canada

Abstract. Autonomous cyber agents may be developed by applying reinforcement and deep reinforcement learning (RL/DRL), where agents are trained in a representative environment. The training environment must simulate with high-fidelity the network Cyber Operations (CyOp) that the agent aims to explore. Given the complexity of network CyOps, a good simulator is difficult to achieve. This work presents a systematic solution to automatically generate a high-fidelity simulator in the Cyber Gym for Intelligent Learning (CyGIL). Through representation learning and continuous learning, CyGIL provides a unified CyOp training environment where an emulated CyGIL-E automatically generates a simulated CyGIL-S. The simulator generation is integrated with the agent training process to further reduce the required agent training time. The agent trained in CyGIL-S is transferrable directly to CyGIL-E showing full transferability to the emulated "real" network. Experimental results are presented to demonstrate the CyGIL training performance. Enabling offline RL, the CyGIL solution presents a promising direction towards sim-to-real for leveraging RL agents in real-world cyber networks.

Keywords: cyber network operations · RL training environment · deep reinforcement learning · sim-to-real agent training and transfer

1 Introduction

The modern world relies heavily on the correct operation of communication networks in general and the Internet in particular. Rogue actors regularly attesmpt to disrupt such networks, comprising confidentiality, availability, and/or the integrity of essential information. Organizations and governments are therefore interested in hardening their systems, both by learning about possible attack vectors, and deploying suitable defenses. This is typically carried out as contests between a "red team" attacking the network and a "blue team" defending it, assembling a "game" between the red and the blue teams.

A key challenge is the scalability of the expert red and blue teams. Even though cyber defense and attack emulation tools provide the red and blue agents tools to automate the

É. Renault and P. Mühlethaler (Eds.): MLN 2022, LNCS 13767, pp. 130–144, 2023.
https://doi.org/10.1007/978-3-031-36183-8_9

IT workflow by enabling the staging, scripting commands and filling in the payloads, the sequential decision making at each action step still relies on the human cyber expert.

Deep Reinforcement Learning (DRL) enabled autonomous agents are envisioned to carry out network cyber operations (CyOps) with superior decision-making capabilities. A growing body of work is exploring DRL agents for use cases from autonomous cyber penetration tests to network red team and blue team exercises [1–6]. Similar to other real-world applications such as self-driving vehicles and autonomous robots, the DRL algorithms may train the agents to learn and optimize their Course of Actions (CoAs) for multi-stage operations in the complex and dynamic environment of cyber networks.

Developing applicable agents in real cyber networks requires first a DRL agent training environment that models the cyber network where the agent will operate. The CyOp agent training environment has started drawing interest in the community. A few environments have been reported, open-sourced, or even offered as public challenges for training CyOp agents, e.g., training a blue defense agent against various red adversary agents [7–13].

A DRL training environment is often a simulator of the real environment. A simulator of the network CyOp environment is particularly challenging, given the complexity of modern networks and CyOp actions. Capturing configurations of and interactions across numerous network components, the simulator is prone to be incomplete, as it needs to represent the host and network states which are in fact unknown at times, even to an expert. The state changes caused by actions are stochastic instead of being deterministic. Changes in network configurations and in the involved CyOp action also often bring about code redesign of the simulator.

To manage this problem, current CyOp simulators [7, 8, 10–12] are built on abstractions of actions and simplification of states. The agent trained from such a simulator however cannot be transferred or deployed to the real network, as the actions learned and state data used in training deviate from reality. For example, the real red team may apply tools to select from more than 10 different network discovery actions, each relying on different techniques with their corresponding network configurations. The red agent on the other hand is trained in the simulator on one abstract action of "network discovery" [10–13]. As a result, the trained agent is not transferrable to a real network.

CyOps grounded in realistic cyber networks share the same problem with many real-world RL applications: a good simulator is essential but hard to build [14, 15]. To attain high fidelity, real system data may be used. However, gathering data from the real cyber network is equally time-consuming. Although the RL training environment for achieving sim-to-real has been well investigated and advanced in other domains such as robotics, for example, through using environment images directly [16], the solutions are not applicable to cyber networks.

This work investigates approaches for building the CyOp training environment towards the goal of sim-to-real agent training and transfer. A Cyber Gym for Intelligent Learning (CyGIL) is presented, which is a unified deployment across both the real (or emulated) CyOp network, namely CyGIL-E, and its mirroring simulator, namely CyGIL-S. CyGIL-E runs on the real network or its emulated version over virtualized hardware of VMs (Virtual Machine). Actions in CyGIL-E use operational tools as they are used in the real network. CyGIL-S is auto-generated using CyGIL-E data to mirror the real environment. To our knowledge, this is the first CyOp training environment uni-fied on both real (emulated) and simulated cyber networks with a complete cross-training and evaluation loop. The contributions include the following:

1. A cyber network RL environment that supports efficient agent training with high fidelity in a unified emulator (or real network) and simulator
2. Unsupervised auto-generation of CyGIL-S (the simulator) from the real (or emulated) network
3. A unified DRL training framework across CyGIL-E and CyGIL-S, demonstrating effective representation learning to reduce the time for data collection and for agent training

The rest of the paper is organized as follows. Section 2 presents the CyGIL system including both CyGIL-E (emulation-based) and CyGIL-S (simulation-based) training environment. Section 3 elaborates on the issues in generating CyGIL-S through experi-mental results. Section 4 presents the unified agent cross-training and evaluation solution. Section 5 draws concluding remarks.

2 Unified CyGIL-E and CyGIL-S

2.1 System Design

CyOps involve sequences of actions and their impact on network states over a short or long period. In the process, the attacker, referred to as the red agent, takes a sequence of actions to form and complete a cyber-kill chain [17] to break the confidentiality, integrity and availability of the network information and services. Meanwhile, the defender, referred to as the blue agent, must sustain the network mission objectives, throttling the kill chain, removing the red relics and recovering the compromised functionality. Red and blue agents use their respective operation tools to conduct the described CyOps.

The conceptual framework of the unified CyGIL is shown in Fig. 1 (with some of the notations introduced more formally in Sect. 2.2). Both red and blue agents choose, using appropriate tools, specific actions that are applied to the training environment. The training environment knows about the game objectives and returns both observations (success or failure of the chosen action, information gained from successful execution of an action, etc.) as well as a reward. The reward reflects how well the action advances the agents' objectives.

An implementation of the unified CyGIl-E and CyGIL-S is illustrated in Fig. 2. The CyGIL network can be either a real network or its emulation on virtualized hardware, encompassing network assets including the CyOp tools, and additionally actors such as users. Figure 2 presents the implementation details of the mini CyGIL configuration for research, where the network is emulated on virtualized Mininet switches using the Open Network Operating System (ONOS) Software Defined Network (SDN) controller. Large networks are emulated using vSphere [18].

In the CyGIL training node, the CyGIL environment (env) library wraps the network and the training game into a CyGIL env python class to stand up the gym instance that provides the openAI gym interface [19] to the agent(s) for DRL training. Each gym

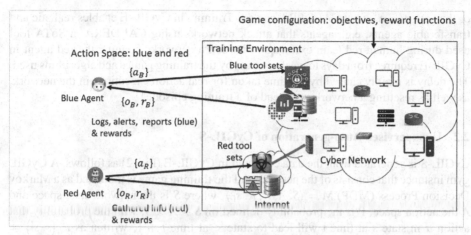

Fig. 1. Modeling CyOp environment to CyGIL framework: action $a_B \in A_B$, $a_R \in A_R$; observation $o_B \in O_B$, $o_R \in O_R$; reward r_B and r_R produced by R_B and R_R respectively (Color figure online)

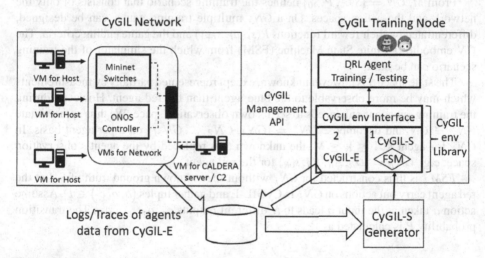

Fig. 2. CyGIL system implementation: 1 – agent training; 2 – representation learning, agent transfer and verification; dashed lines – the interface between CyGIL library and the real (or emulated) network and CyOp tools; C2: Command and Control of CyOp toolset(s)

training instance consists of the network and the training game. A training session may open its CyGIL gym instance in either CyGIL-E or CyGIL-S.

In [9], it is demonstrated that red agents in CyGIL-E can learn and optimize their decision engines to achieve different attack objectives across the network. The agent learned, using the CALDERA red team tool [20] shown in Fig. 2, to form an optimized end-to-end kill chain step-by-step, from network discovery, command-and-control, credential access, privilege escalation, defense evasion, lateral movement, to information collection and exfiltration, all starting from knowing nothing about either the network or the actions, i.e., what CALDERA can do. Training in CyGIL-E enables realistic and transferable agents, e.g., agents that attack networks using CALDERA, a SoTA tool used during human red team exercises to harden a network. Training the red agent in CyGIL-E requires from days to weeks, varied by the training games and algorithms used. The delay is mainly caused by the time taken for real action executions in the network, as well as resetting a network at the end of a training episode.

2.2 Unsupervised Auto-generation of CyGIL-S

CyGIL-S is generated from the data collected in CyGIL-E (Fig. 2) as follows. A CyGIL gym instance that consists of the network and the training game is modelled as a Markov Decision Process (MDP) $M = \langle S, A, P, R, s_0 \rangle$, where S is the network state space and A the action space. P is the probability defined on $S \times A \times S$, with the probability that action a in state s at time t will lead to state s' at time $t + 1$, written as $P_a(s, s') = \Pr(s_{t+1} = s'|s_t = s, a_t = a), a \in A, s, s' \in S$. R is the reward function defined on $S \times A \times S \to \mathbb{R}$, with $R_a(s, s')$ as the reward function after transitioning from state s to state s', due to action a. The initial distribution of S is s_0.

From M, $GN = \{S, A, P, s_0\}$ defines the training scenario that consists of only the network and the action spaces. On a GN, multiple training games can be designed, differentiated by their reward functions $R(s_t, a_t, s_{t+1})$ and the game ending criteria. The GN embodies a Finite State Machine (FSM) from which the simulator of the training scenario can be built.

The state space S is however unknown, except represented partially by measurements which may be more observable to the blue agent than the red agent. However, during the training, an agent only needs to see its own observation space. We thus approximate $GN^* \cong GN$ and decompose $GN^* = \{GN_1, GN_2, \ldots GN_M\}$ on a per-agent basis. In GN_k for agent k, $1 \leq k \leq M$, the unknown S is replaced by the agent's observation space, e.g., $GN_R = \{O_R, A_R, P_R, s_{0R}\}$ for the red agent.

FSM_k is thus constructed for GN_k without the unknown ground truth of S. Let the red agent carry out actions on GN_R in CyGIL-E and gather tuples $(a, o, o') \in \mathcal{D}$. Assume action a taken at the input o leads to N different outputs $o'_1, o'_2, \ldots, o'_N$. The transition probability P is calculated as

$$P_a(o, o'_i) = \frac{C^{o'_i}_{(a,o)}}{\sum_{j=1}^N C^{o'_j}_{(a,o)}}, \sum_{j=1}^N P_a(o, o'_j) = 1 \tag{1}$$

where $P_a(o, o'_i) = \Pr(o_{t+1} = o'_i|o_t = o, a_t = a)$, and $C^{o'_i}_{(a,o)}$ counts the output observation o'_i when action a is taken at the input observation $o_t = o$. In the case where multiple agents take actions at each action step, a extends to a vector to account for the different actions taken by agents. The details are not included here for brevity.

The CyGIL-S generated in this way produces observation data o'_i upon receiving (a, o), forming the transition (a, o, o'_i) according to $P_a(o, o'_i)$ for agent training. This enables a lightweight CyGIL-S that supports fast agent training, as shown in the next section. The CyGIL-S generation is data-centric, agnostic to network topology, action sets, any game parameters and any values of the data. The CyGIL-S generator code is therefore reusable for different M and GN when the network and/or game changes.

3   CYGIL-S Evaluation

3.1   Test Data Collection Time

A key challenge is to collect, in as little time as possible, a data set D that can generate a sufficient CyGIL-S. A sufficient CyGIL-S embeds all required state space and transition probabilities to train the agent to the optimal policy. Data collection in the real-world physical system is time-consuming and expensive. Therefore, identifying the sufficient set D is critical.

An example is used here to illustrate the time expense issue. The experiment network is depicted in Fig. 3. All hosts inside the network can reach the Active Directory Server and its Domain Controller (DC) on host 6 which is a Windows 2016 server. Other hosts in the network are Windows 10 machines except for hosts 1 and 9 which run on Linux Ubuntu 18. Hosts 1 and 2 are reachable from the external "Internet" where the red agent is trained to operate the attacker's C2. Hosts on the same switch belong to the same subnet and can communicate with each other. Between different subnets, firewall rules controlled by ONOS allow host 5 to communicate with hosts 2 and 3 in addition to hosts in its subnet. Each host communicates with some other hosts at any given time, forming the user traffic.

The training game defines the agent's objective as to land on the DC of ADS on host 6. If it succeeds, the red agent will have the admin privilege to breach the entire domain. The red agent's action space is shown in Fig. 4, encompassing key tactic groups in the ATT&CK framework. These actions mimic the general tactics of several Advanced Persistent Threat (APT) groups [17] to enable the network end-to-end kill chain. As the initial state of the training game, the red agent has already compromised host 2 via phishing and implanted a "hand", i.e., malware. The hand reports back to the agent at C2, as supported by the CALDERA staging framework.

The reward function R for each action step is defined as $R = G - L$ where G is the gain and L is the cost. The game sets $G = 0$ for any action unless the agent reaches the objective, where it receives $G = 100$. It also sets $L = 1$ and $L = 8$ per hand, for an action that has all the required input parameters and an action that does not, respectively. This assigns a higher penalty to the selection of actions which do not even have the required parameters to form the execution commands. The reward after each action step is thus always negative unless the objective is reached. The game ends when either the agent achieves the objective or the game reaches the maximum number of steps which is set to 80. When failing to achieve the objective, the red agent ends up with a negative accumulated return of -80 or lower. The most optimized CoA can reach the objective in 6 steps, with some steps requiring concurrent actions taken by multiple hands. The red agents achieves an accumulated reward of 92 in this case, provided the random action outcomes all favor the agent.

Emulating the network on a Windows laptop that runs on Intel(R) Core (TM) i9 and 64 GB RAM, the agent is trained using CyGIL-E on a second laptop that has an Intel(R) Core (TM) i7 and 64 GB RAM. The time to reach the optimized policy ranges from 7

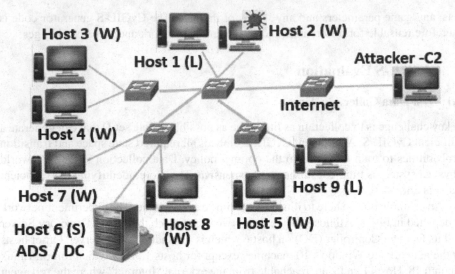

Fig. 3. The Example Network (Color figure online)

ATT&CK Tactics	ATT&CK Techniques	Game action ID
Discovery	T1135 – Network Share Discovery	0
Discovery	T1087.002 - Enumerate AD user accounts	11
Discovery	T1018: Enumerate Active Directory computer objects	12
Discovery	T1016: collect ARP details	13
Discovery	T1003: reverse nslookup	14
Credential Access	T1003.001: Mimikatz to extract credentials	7
Credential Access	T1110.001: Brute Force credentials of domain user (NTLM or Kerberos)	10
Privilege Escalation	T1548.002: Download and execute Sandcat file agent as admin user	8
Lateral Movement	T1021.006: Execute Sandcat from fileshare remotely with WinRM	1
Lateral Movement	T1021.006: Execute Sandcat from fileshare remotely using PsExec	2
Lateral Movement	T1021.006: Copy Sandcat File with SCP and execute using PsExec	3
Lateral Movement	T1021.006 Execute Sandcat File remotely from system using WinRM	4
Lateral Movement	T1021.006 Mimikatz PSH and PsExec for launch Sandcat file on remote machine	9
Lateral Movement	T1021.006: PsExec to copy and launch Sandcat file on remote machine	15
Lateral Movement	T1570: Copy Sandcat file to remote system using WinRM and SCP	6
Lateral Movement	T1570: Copy Sandcat file to file share	5

Fig. 4. Agent Action Space – key TTPs from ATT&CK framework

to 20 *days*, depending on the learning algorithms used [9]. Using data collected from CyGIL-E to generate CyGIL-S according to Eq. (1) and configuring the same game on CyGIL-S to train the agent afresh, the DQN [21], PPO [22] and C51 [3] Rainbow agent can all learn the optimal policy with average training times of 17.31, 25.92 and 5.76 min respectively, much faster than in CyGIL-E.

Different data sets D from CyGIL-E are tested in generating CyGIL-S. From Table 1, to generate a sufficient CyGIL-S that can train the agent to the optimal policy, the time

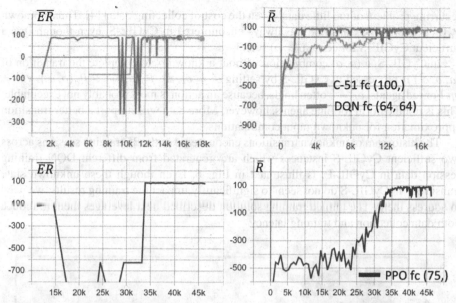

Fig. 5. Agent training in CyGIL-S - X axis: training steps; $\overline{R}$: Average training reward; $\overline{ER}$: Average evaluation rewards, fc: the architecture of the fully connected layer

required approximates that for training the agent in CyGIL-E. This time is unsatisfactorily long. Although algorithm parameter tuning, tests of different algorithms and agent training with new reward functions and game–ending criteria can be carried out on CyGIL-S as shown in Fig. 5, which are not feasible on CyGIL-E, reducing the data collection time is desirable. This is discussed in the next section.

Table 1. Data Set for Generating CyGIL-S

Source of Data Set D	Sufficient CyGIL-S Generated
Random Plays of 10 days	No
A DQN training session	Yes
A PPO training session	Yes

3.2 Unknown Transitions in CyGIL-S

Both training and evaluation in CyGIL-S experience unknown states and action pairs. This is because the agent in CyGIL-S may step into the states that have not been reached in CyGIL-E, nor embedded in the data set D collected, given the large state space. The data-based simulated RL training environments in every real-world application face the recurring problem of too little data, given the cost of collecting data [14]. Thus unknown $P_a(o, o')$ for some transition are always encountered in CyGIL-S, even though it is a sufficient CyGIL-S for certain (a, o) sets.

In CyGIL-S, a (a, o) combination without a known o' to complete the transition of (a, o, o') is processed in CyGIL-S by setting $o' = o_I$, s.t. $P_a(o, o_I) > P_a(o, o'), \forall o' \neq o_I$. For red agents, $o_I = o$ often holds, because an action is most probably not executable. This is not the norm for blue agents, however. More data will support a better transition approximation for unknown input combinations.

The histogram of unknown transitions encountered in the 25 training sessions across two sufficient CyGIL-S instances which are generated from different DQN training session data in CyGIL-E, is illustrated in Fig. 6. Even though these unknown state transitions in CyGIL-S do not seem to significantly impact the training results, we need to address them. The unified training solution described next leverages them to reduce both data collection and training latency.

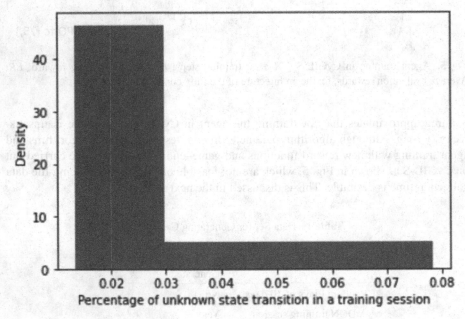

Fig. 6. Unknown transition distribution in sufficient CyGIL-S

4   Unified CyGIL Training

4.1   The Cross Training Loop

To reduce the required time for data collection in CyGIL-E and the overall agent training time, a unified CyGIL-E and CyGIL-S solution is developed. Its mechanism consists of a closed loop of transfer and continuous learning, as illustrated in Fig. 7. The same experiment example (Fig. 3 and Fig. 4) is used to illustrate the details.

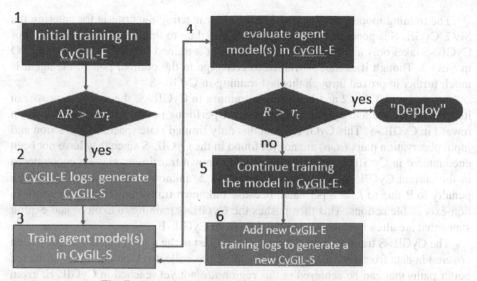

Fig. 7. Agent training in unified CyGIL-E and CyGIL-S

As shown in Fig. 7, the unified training loop starts in Segment (SEG) 1 by training the agent in CyGIL-E until the reward improvement jumps over a threshold, $\Delta R > \Delta r_t$. At this stage, the training average reward is often far below the optimal value and the evaluation reward does not yet show any improvement. However, some good paths have already been traversed in the state space. This training session is used as a representation learning to move towards the better region(s) of the state space. Using the above network and game example, a sample efficient algorithm DQN is used in this SEG, as shown in Table 2. After 113 training episodes with more than 8k steps, the training average reward improved to -0.9 compared to the initial value of -912, with the episode length (the number of steps it takes the red agent to either successfully complete an episode or reach the end of an episode training) reducing from 80 to 24. The average performance of the subsequent training episodes in CyGIL-E degraded as shown in Fig. 8 (a) and (b), as expected, as the model is far from being converged.

Table 2. Training Results across CyGIL-E & CyGIL-S

Loop SEG & Model	In CyGIL-E or S	T. Elapsed Time	Training Average Reward and Best Episode Length
1 - DQN	CyGIL-E	35.5 h	−0.9, 24
3 - PPO	CyGIL-S	29 m	26.5, 10
5 - PPO	CyGIL-E	1.1 h	26.5, 10
3 - C51	CyGIL-S	4 m	92, 8

The training loop moves to SEG 2, trigged by meeting the criteria for entering the SEG. CyGIL-S is generated from the data collected up to this point. The generation of CyGIL-S takes only a few seconds. Then the agent is trained in the CyGIL-S using PPO in SEG 3. Though it cannot be trained to converge to the optimal policy, the agent is much further improved through this fast training in CyGIL-S.

As shown in Table 2 and Fig. 8(c), in training in CyGIL-S, the model improves in its training reward and episode length and yet performs poorly in average evaluation reward in CyGIL-S. This CyGIL-S contains only limited state space. Many action and input observation pairs (a, o) are not yet found in the CyGIL-S since they have not been encountered in CyGIL-E yet. That is, many unknown transitions may be encountered in the current CyGIL-S. As described in Sect. 3.2, unknown transitions cause a high penalty to R due to $L = 8$ per hand, because unknown transitions frequently result in non-executable actions. This then pushes the CyGIL-S training to explore and exploit states that are already embedded in the data from CyGIL-E.

The CyGIL-S training indeed wants to move fast in the region that has already been covered in data from CyGIL-E. While CyGIL-E has collected the data for a region, the better paths that can be achieved in this region are not yet reached in CyGIL-E, given the slow action execution in CyGIL-E. It should be noted many more paths are in the data from CyGIL-E than the paths that have been executed in CyGIL-E, because many new paths can be formed by concatenating the state transitions in the data. Stepping through paths is very slow in CyGIL-E. The CyGIL-S training compensates for the latency problem by quickly exploring and exploiting the potential paths to generate a better model.

When its policy cannot be improved further in the current CyGIL-S, the agent is transferred to CyGIL-E as shown in SEG 4. In CyGIL-E, if the agent model already exceeds the required return (r_t), the agent training is completed and the agent can be deployed in the real (emulated) network. Otherwise, the agent model continues its training in CyGIL-E as shown in SEG 5, this time leveraging the knowledge it obtained from training in CyGIL-S. This quickly leads the agent to explore regions with much higher returns than during its previous training session in CyGIL-E.

Moving from SEG 5 to SEG 6, i.e., returning to continued training in CyGIL-S, is triggered by counting the number of training episodes CyGIL-E. After every 4 episodes, the newly collected data is added to the previously collected logs to generate a new CyGIL-S as shown in SEG 6. Using this new CyGIL-S, training loops back to SEG 3 for agent training in CyGIL-S while the training in CyGIL-E also continues in parallel to collect more data from further training episodes. Again, in CyGIL-S, the agent model is trained from scratch without using the model from CyGIL-E, given the fast training in simulation. For this training game scenario, after collecting data from an additional 8 episodes in CyGIL-E, the new CyGIL-S trains successfully an optimized agent policy using the C51 rainbow algorithm [19] (Table 2 and Fig. 8(d)). This trained model achieves the optimal CoA when transferred and evaluated in CyGIL-E, in all 50 evaluation runs.

It is noted that when the last 8 episodes were executed within CyGIL-E, the model training in CyGIL-E was still far from converging to the optimal policy. The average episode length in CyGIL-E already improved to around 10. However, the training still requires a long time before approaching the optimal policy. The reduced learning rate, the

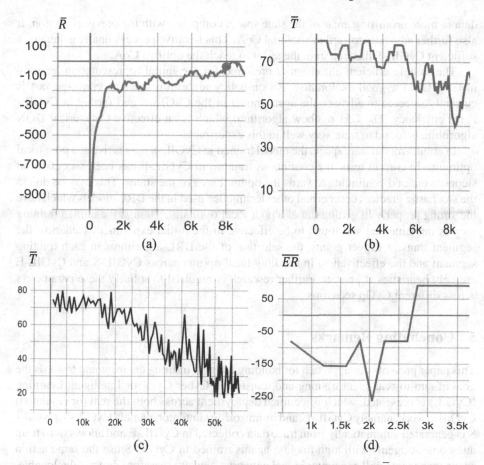

Fig. 8. Unified training across CyGIL-E and CyGIL-S - X axis: training steps; $\bar{R}$: Average training reward; $\bar{T}$: Average episode length; $\overline{ER}$: Average evaluation rewards; (a) and (b) Initial SEG 1 in CyGIL-E using DQN algorithm; (c) First training in CyGIL-S using PPO algorithm (SEG 3); (d) Second training in a new CyGIL-S using the C51 rainbow algorithm (SEG3)

selected "good" actions which are always executable, and the added evaluation episodes all extend the time required in CyGIL-E. Yet, the collected data already embed the best CoA and enable CyGIL-S to train the agent to the optimal policy.

4.2 Discussions

The agent training in CyGIL-E can be considered a form of representation learning to find the right set of data to collect, which embodies the training path towards the optimal model. This is much more efficient than collecting random data from a real network or emulated network without training the agent. The CyGIL-S training rapidly pushes the agent model towards the better areas in the region, towards the optimal CoA. The model trained in CyGIL-S is transferred and continuously trained in CyGIL-E to collect further data in more promising areas of the state space, compared with the previous session. It also further advances toward the optimal CoA. This iterative process finally generates a sufficient CyGIL-S that can train the agent to reach the optimal CoA.

The sample efficient algorithm is preferred for the initial representation learning in CyGIL-E for good exploration. An on-policy algorithm that can converge fast is currently selected to advance the agent model in the CyGIL-S and further in CyGIL-E for efficiency. The C51 rainbow algorithm, which is an effective Categorical DQN algorithm, is found to train very well in this experiment.

In amulti-modal state space, the model trained in CyGIL-S may arrive at a poor local optimum. In current tests, the continuous training in CyGIL-E has been successful in stepping out and continuing towards the optimal area of the states. This may be due to the stochastic gradient descent and other techniques used in the DRL models which have the strong property in getting out of a poor local optimum. Though the unified training loop is implemented and found to be effective in the initial experiment scenarios, the segment transfer trigger points, the selection of the DRL algorithms in each training segment and the effectiveness in handling local optima across CyGIL-S and CyGIL-E are only heuristics at present. Further research is required to solidify these parameters across different CyOp scenarios.

5 Concluding Remarks

This paper presents our approach for building a CyOp training environment towards the goal of sim-to-real agent training and transfer. A Cyber Gym for Intelligent Learning (CyGIL) is presented, which is a unified deployment across both the real (or emulated) CyOp network, namely CyGIL-E, and its mirroring simulator, namely CyGIL-S. CyGIL-S is generated automatically from trace data collected in CyGIL-E and allows us to train autonomous agents with high fidelity: agents trained in CyGIL-S use the same action space that agents will encounter in real networks, and are therefore directly deployable, unlike other CyOp training environments based on simulation. Training agents in CyGIL-S takes only a fraction of the time it would take to train these agents in the real/emulated network and allows us to explore various "what-if" scenarios that would otherwise be infeasible.

A challenge in building CyGIL-S is to identify how much data needs to be collected from CyGIL-E to build a sufficient simulator (i.e., a simulator that allows a trained agent to discover the optimal course of action). In the initial version, as summarized in Table 1 at the end of Sect. 3.1, we would require CyGIL-E to complete the training of a single agent to collect enough data to build a sufficient CyGIL-S. While this may seem unattractive, once this CyGIL-S is constructed, we can then use it to explore additional training algorithms, modify game objectives, etc., with little additional runtime cost.

A key challenge in collecting fewer data is the occurrence of unknown state transitions: the FSM underlying CyGIL-S depends on having observed state (or observation) changes as a consequence of actions taken by an agent in the real or emulated network. For the sufficient CyGIL-S described in Sect. 3, only a small number of unknown state transitions are encountered during training and they do not prevent the trained agent from learning the optimal course of action. However, reducing the collected data will increase the number of these unobserved state transitions, so any reductions have to carefully manage this problem.

Our solution, described in Sect. 4, uses a unified training approach combining both CyGIL-E and CyGIL-S. In a nutshell, we collect initial data to build a first CyGIL-S. An agent trained with this incomplete simulator will move relatively quickly towards more promising courses of actions. Transferring the agent back into CyGIL-E, we can collect more relevant state transitions as the more knowledgeable agent explores more promising regions of the state space. With this collected data, we can then build better versions of CyGIL-S, and in our running example, the trained agent learns the optimal course of action after one iteration through the loop. Overall, as summarized in Table 2, this speeds up agent training compared to training an agent purely in CyGIL-E. It also significantly reduces the time to build a sufficient CyGIL-S.

The approach presented in this paper seems promising. Future work will expand on the key iterative loop shown in Fig. 7: what parameters will trigger switches between CyGIL-E and CyGIL-S, what training algorithms to best use in different stages of the training cycle, etc. We will also explore additional scenarios to explore in more depth how robust this approach is against getting stuck in local optima.

References

1. Chaudhary, S., O'Brien, A., Xu, S.: Automated post-breach penetration testing through reinforcement learning. In: Proceedings of 2020 IEEE Conference on Communications and Network Security (CNS) (2020)
2. Ghanem, M.C., Chen, T.M.: Reinforcement learning for intelligent penetration testing. In: Proceedings of Second World Conference on Smart Trends in Systems, Security and Sustainability (WorldS4), pp. 185–192 (2018)
3. Nguyen, H., Nguyen, H.N., Uehara, T.: Multiple level action embedding for penetration testing. In: The 4th International Conference on Future Networks and Distributed Systems (ICFNDS) (2020)
4. Zennaro, F.M., Erdodi, L.: Modeling penetration testing with reinforcement learning using capture-the-flag challenges and tabular Q-learning. arXiv preprint arXiv:2005.12632 (2020)
5. Schwartz, J., Kurniawati, H.: Autonomous penetration testing using reinforcement learning. arXiv preprint arXiv:1905.05965 (2019)
6. Sutana, M., Taylor, A., Li, L.: Autonomous network cyber offence strategy through deep reinforcement learning. In: Proceedings of SPIE conference on Defences and Commercial Sensing, April 2021 (2021)
7. Baillie, C., Standen, M., Schwartz, J., Docking, M., Bowman, D., Kim, J.: CybORG: An autonomous cyber operations research gym. arXiv preprint arXiv:2002.10667 [cs], 2 (2020)
8. Molina-Markham, A., Miniter, C., Powell, B., Ridley, A.: Network environment design for autonomous cyberdefense. arXiv preprint arXiv:2103.07583 (2021)
9. Li, L., Fayad, R., Taylor, A.: CyGIL: A cyber gym for training autonomous agents over emulated network systems. arXiv preprint arXiv:2109.03331 (2021)
10. Schwartz, J., Kurniawatti, H.: NASim: Network Attack Simulator (2019)
11. Microsoft. CyberBattleSim Project - Document and source code, GitHub (2021)
12. Standen, M., Lucas, M., Bowman, D., Richer, T.J., Kim, J., Marriott, D.: CybORG: A gym for the development of autonomous cyber agents. arXiv preprint arXiv:2108.09118 (2021)
13. TTCP CAGE Challenges, GitHub - cage-challenge/cage-challenge-2: TTCP CAGE Challenge 2

14. Dulac-Arnold, G., Mankowitz, D., Hester, T.: Challenges of real-world reinforcement learning (2019)
15. Nair, A., Dalal, M., Gupta, A., Levine, S.: Accelerating online reinforcement learning with offline datasets. arXiv preprint arXiv:2006.09359 (2020)
16. Peng, X.B., Andrychowicz, M., Zaremba, W., Abbeel, P.: Sim-to-real transfer of robotic control with dynamics randomization. In: 2018 IEEE International Conference on Robotics and Automation (ICRA), Brisbane, Australia, May 2018 (2018)
17. MITRE Corp. MITRE ATT&CK knowledge base (2021)
18. VMWare Vsphere Documentation. https://docs.vmware.com/en/VMware-vSphere/index.html
19. OpenAI, Gym Documentation. https://www.gymlibrary.dev (2022)
20. MITRE Corp. CALDERA - Document and source code. GitHub (2021)
21. Farebrother, J., Machado, M.C., Bowling, M.: Generalization and regularization in DQN. arXiv preprint arXiv:1810.00123 (2018)
22. Schulman, J., Wolski, F., Dhariwal, P., Radford, A., Klimov, O.: Proximal policy optimization algorithms. arXiv preprint arXiv:1707.06347 (2017)
23. Bellemare, M.G., Dabney, W., Munos, R.: A distributional perspective on reinforcement learning. In: International Conference on Machine Learning (2017)

Deep Learning Based Camera Switching for Sports Broadcasting

Hamid Reza Tohidypour[✉], Yixiao Wang, Mahsa T. Pourazad, Panos Nasiopoulos, Gurpreet Heir, Derinsola Ibikunle, Anthony Li, Fawaz Ahmed Saleem, and Zhaobang Luo

Department of Electrical and Computer Engineering, University of British Columbia, Vancouver, Canada
{htohidyp,yixiaow,pourazad,panosn}@ece.ubc.ca

Abstract. Switching camera views when broadcasting sport games is essential for improved quality of viewing experience. Traditionally, expensive equipment and a broadcast director are employed to choose the optimal camera view during the game. In this study, we propose a novel deep learning based method to automatically switch between cameras based on importance of the scene detected in each view. Here, in order to show the validity of our approach, we chose to train our network for ice hockey, as the network needs to be retrained for each sport, using a dataset that corresponds to the specific game. Our method uses a YOLOv4 model that accurately detects the important objects for ice hockey, namely players, puck, net, goalie, and referee. Our novel camera switching view scheme uses the confidence values of the detected objects and temporal tracking to choose the most important camera view for that instance.

Keywords: Multi-camera Switching · Object Detection · YOLOv4 · Convolutional Neural Network

1 Introduction

Live broadcasting of sports had a major impact on the way sporting events are viewed. The use of multiple cameras with different capabilities, covering a variety of viewing angles, allows live broadcasting to offer a unique perspective to home viewers or the spectators on the stands through large stadium displays. Traditionally, such broadcasting normally requires expensive equipment and control rooms with directors and large teams that have advanced skills and expertise. Directing active cameras at the position of the players and switching between the cameras according to action and interest of the fans is exceptionally challenging. In a sense, live sports broadcasting is a storytelling process, like any other production, making sure that the viewers enjoy the best quality of experience during the game [1]. Several attempts have been made to replace or assist directors and crew in live sport broadcasting, ranging from traditional machine learning to deep learning.

É. Renault and P. Mühlethaler (Eds.): MLN 2022, LNCS 13767, pp. 145–152, 2023.
https://doi.org/10.1007/978-3-031-36183-8_10

In [2], an automatic director assistant was proposed for field hockey that achieves accuracy of 51.59% for camera switching. This method focuses on the players rather than ball (puck), thus priority is given to the number of players on the scene and not scoring. In [3], a deep learning based directing system was proposed for live broadcasting of football matches. This approach is event based, detecting instances of shooting, player falling, goal kick, thrown in, and corner and free kicks. The network learns to choose a camera using feedback from the detected action, introducing a 30 s delay. For such an approach to be successfully implemented, we need a huge dataset that covers an immense number of events and actions, as they are almost endless possibilities in the game of football. The method described in [4] is designed for ice hockey and performs an automatic camera selection using the Hidden Markov Model to create personalized video programs for users that are more interested in the performance or positions of the players from different perspectives than the game it-self. In that respect, this method was player-centered rather than puck-centered or action-centered. In [5], an action-centered automatic camera switching method is proposed for ice hockey games that uses Faster-RCNN [6] to detect the puck, players, and net. This method achieved an object detection accuracy of 78.9% and camera switching accuracy of 75%, indicating that there is still room for improvement.

In this paper, we propose an innovative camera switching method which is based on deep learning and a temporal tracking scheme to automatically pick the most important view to be broadcasted. Here, in order to show the validity of our approach, we chose to train our network for ice hockey, as the network needs to be retrained for each sport, using a dataset that corresponds to the specific game. We based our method on the YOLOv4 architecture [7]. Our model receives video feeds from all the cameras around an ice hockey arena and detects the puck, net, goalie, players, and referees in real time with very good precision. A novel camera switching algorithm that weights the objects detected by each camera view according to their importance and uses the predicted confidence values for the different objects and temporal tracking to choose which camera view to be broadcasted. Our results showed that our proposed method predicted the best views with 98% accuracy.

The rest of this paper is organized as follows. In Sect. 2, we present our method, explain our dataset, the labeling approach, and the network architecture we used. Section 3 presents the performance evaluation of our method and discusses the results. Finally, Sect. 4 concludes our paper.

2 Our Approach

Sports broadcasting involves the use of several cameras located at different places of the stadium. For the specific case of ice hockey, some of the camera views cover the long sides of the arena, giving a wider view of the field. These camera views are called side views. The other cameras are mounted behind the nets to capture the action near the net, and to give a closer view to scoring opportunities. These camera views are called goalie views. Figure 1 shows an example of a multi-camera setup for hockey arenas. In this paper we propose a method that processes live video feeds form all the cameras and predicts when to switch from one view to another view. In this regard, first we propose a deep learning-based object recognition approach that receives the videos from all the views to detect the players, puck, net, goalie, referees. Then, we use this information and some conditions to decide which view to be broadcasted. Details about the dataset we used to train network, the architecture of our network, and the scheme and policies we introduced for camera view switching are presented in the following subsections.

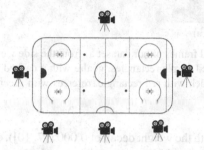

Fig. 1. An example of a multi-camera setup for ice hockey arenas.

2.1 Data Collection and Labeling

In order to generate a comprehensive dataset for our task, first we collected a large number of videos from local amateur ice hockey games. These videos were captured by cameras mounted on the side views and goalie views. In addition, we also collected a large number of professional ice hockey videos from YouTube [8]. From all the videos we selected 5000 well representative frames for the training-validation phase, avoiding subsequent, similar frames and preferring frames that included the puck. Almost half of these frames were selected from the videos captured from the side view, and the remaining from the goalie view. The choice of the frames included in our dataset was based on the view choices made by professional directors covering real-life hockey games. In our approach, the objects of interest are the players, goalie, puck, net, and referee. The presence of these five types of objects which is translated into a type of action is used to determine the best camera view for each instance. Example of labeled frame from the goalie view and side view which are used for training are shown in Fig. 2(a) and (b), respectively.

2.2 Our Deep Learning Network

We chose a state-of-the-art object recognition approach named YOLOv4 [7], from the YOLO family [9] as our object recognition architecture. YOLOv4 reported to give promising results in detecting small objects with very fast inference time [7]. This is important since one of the main drawbacks of the Faster-RCNN based camera switching method proposed in [5], was the low accuracy of detecting the puck. In our implementation we used the YOLOv4 darknet framework found in [10]. We increased the default input size to 608x608 in order to be able to detect small objects (i.e., the puck), as smaller sizes would reduce the accuracy detection of the network. We set the batch size to 64, the

a)   b)

Fig. 2. Examples of labeled frames of our dataset a) from the side view b) from the goalie view, showing players with dashed purple rectangles, goalie with a dashed cyan rectangle, net with a dashed yellow rectangle, puck with red rectangle, referees with green rectangles. (Color figure online)

learning rate to 0.001 (with the weight decay of 0.0005 [7, 10]), and changed the number of filters in the layers related to the number of classes to 5 – representing the puck, player, goalie, referee and net. 80% of the training-validation dataset was randomly selected as the training dataset and the remaining 20% was considered for the validation phase. The cutmix and mosaic augmentations [10] were disabled since they did not help improve the accuracy for our task. We trained our YOLOv4 model using the Nvidia V100 Volta GPU, with 32 GB memory available on advanced computer clusters [11].

2.3 Camera Switching

For the camera switching scheme, the first step is to detect the players, goalie, net, referee and the puck using our object recognition model. Then, our algorithm considers the position and confidence level of all the detected objects to assign a score for each camera view. However, since each object plays a different role in a given scene, the score influence of each is different. In this respect, in our approach the object with the highest influence is the puck, followed by the goalie, player, net and referee as our empirical studies have shown. The latter involved 1000 unseen frames from the side and goalie views. For this reason, we should ensure that our scoring algorithm is biased towards the objects according to their importance. To this end, we assigned a weight to each object type according to its importance: 20, 2, 1, 1, and −2 for the puck, goalie, net, player, and the referee, respectively. More precisely, the confidence of each detected object in the current camera view is weighted according to its object type and the weighted values are summed up to calculate the score for the current camera view. Note that our method only considers objects with confidence values greater than 20%. The filtered confidence values are converted from the percentage scale to the probability scale. In order to give more importance to the goalie view over the side view when both include players (not necessarily the same), the goalie, the puck and the net, we decided to add 10 to the score of the goalie view.

As explained above, the puck is the most important object for our camera scoring approach. However, detecting the puck correctly can sometimes be a very hard task for the object recognition network. In some cases, the puck can be invisible to the current camera views due to occlusion. For these cases, the camera switching method needs to rely on the previous frames. Thus, we decided to use the information of the three previous frames of the current view to adjust the score for the current view. In this regard, and inspired by the idea of moving average, the score of the current view is updated using the confidence value of the detected puck from the three previous frames before the current frame, as follows:

$$S_{current\ view} = S_{current\ frame} + \sum_{i=1}^{3} \frac{4-i}{4} * 20 * Conf_{puck}(i) \qquad (1)$$

where i indicates the ith frame before the current frame. The maximum value that can be assigned to the i is 3. This value was found using empirical studies. $Conf_{puck}(i)$ represents the confidence value of the detected puck of the ith frame before the current frame.

To avoid fast switching between different cameras which will result in uncomfortable viewing experience and possible flickering, our scheme employs the following two actions: 1) adds 5 to the score of the currently broadcasted, and 2) adds a small delay of 10 frames (one third of a second) before switching again after a recent change of the camera views. Evaluations have shown that the above steps manage to yield the desirable smooth transition between camera views. Figure 3 shows the block diagram of our proposed camera switching scheme.

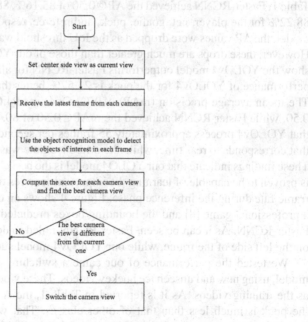

Fig. 3. Our proposed camera switching scheme.

3   Evaluation and Discussion

We compare our method against the state-of-the-art method for camera switching in ice hockey proposed in [5]. To this end, we retrained the Faster-RCNN model presented in [5] using our training and validation dataset to detect the five classes as in our YOLOv4 model, instead of the original 3. The specific Faster-RCNN model uses VGG-16 as the backbone. Note that we used the implementation of Faster-RCNN described in [12], learning rate of 0.01 and batch size of 12, as suggested in [5].

In order to evaluate the performance of our YOLOv4 and the retrained Faster RCNN model, we used our test dataset, which consists of 1500 well-representative frames and examined two different Intersections over Union (IoU) thresholds, 0.50 and 0.75. Table 1 compares the performance of our YOLOv4 model with the Faster- RCNN in terms of average precision (AP) for each of the classes.

We observe that YOLOv4 achieved the (average precision) AP@0.50 of 99.03%, 98.71%, 98.19%, 78.32%, and 99.06% for the player, net, goalie, puck, and referee, respectively. For the IoU threshold of 0.75, YOLOv4 obtained the AP of 92.21%, 97.06%, 85.84%, 26.27%, and 91.83% for the player, net, goalie, puck, and referee, respectively. This shows that the AP values dropped as the IoU threshold was increased from 0.5 to 0.75. In the case of the puck, a huge drop was observed from 78.32% at AP@0.50 to 26.27% at AP@0.75. This is expected due to the relative size of the puck being very small, thus heavily punishing predictions that are slightly off. As it can be seen in Table 1, Faster-RCNN achieved the AP@0.50 of 85.16%, 88.14%, 83.73%, 56.75%, and 88.22% for the player, net, goalie, puck, and referee, respectively. Similar to YOLOv4 model, the AP values were dropped as the IoU threshold was increased from 0.5 to 0.75. However, these drops are much greater than those of our YOLOv4 model. These results show the YOLOv4 model outperforms Faster-RCNN for all 5 categories. The AP@0.50 performance of YOLOv4 for the puck is 21.57% better than that of the Faster-RCNN. The mean average precision (mAP) of YOLOv4 was 94.66% at the IoU threshold of 0.50, while Faster-RCNN achieved the mAP@0.50 of 80.40%. Inference time showed that YOLOv4 process approximately 55 frames per second, much faster than the 30fps that corresponds to real-time, while the inference time was 10fps for the Faster-RCNN. These findings indicate that our YOLO4 model is the best choice for this task. The model is proven to be capable of learning the important features of hockey and keep a suitable frame rate during the inference phase. Figure 4 shows an example of a test frame from a professional game [8] and the bounding boxes predicted by our YOLOv4 model and Faster-RCNN. As it can be seen, Faster-RCNN failed to detect the puck and the player on the left side of the frame, while our YOLOv4 model successfully detected them.

We tested the performance of our camera switching algorithm and the YOLOv4 model, using new and unseen ice hockey videos. These videos are of the same resolution as the training videos. As it is reported in Table 1, the AP@0.50 of our YOLOv4 for the puck is much less than that of other classes. That was the main reason that we decided to introduce temporal tracking through Eq. (1) that uses the information of the previous frames, so our scheme is capable of handling the cases in which the puck is not detected. Similarly, the puck can become invisible due to the occlusion but the temporal tracking approach allows our scheme to avoid unnecessary camera switching. We tested our approach with and without the temporal tracking adjustment for the scores (Eq. 1).

a)   b)

Fig. 4. Example frame selected from the test set [8] with the bounding boxes, labels, and their probabilities that are predicted by a) Faster-RCNN and b) our YOLOv4 model are shown in the images.

Table 1. Detection results of Faster-RCNN and YOLOv4 with different IoU thresholds

Object	Faster-RCNN		YOLOv4	
	AP0.5	AP0.75	AP0.5	AP0.75
Player	85.16%	68.60%	99.03%	92.21%
Net	88.14%	69.56%	98.71%	97.06%
Goalie	83.73%	62.69%	98.19%	85.84%
Puck	56.75%	5.95%	78.32%	26.27%
Referee	88.22%	63.00%	99.06%	91.83%

Our results showed an accuracy of 85% for our camera switching method when temporal tracking is not used. On the other hand, when using temporal information, our method achieved an accuracy of 98% in choosing the most important camera view according to professional directors covering real-life hockey games. In summary, our method, which uses the YOLOv4 model and our camera switching scheme with temporal tracking, outperforms the state-of-the-art approach for camera switching in ice hockey games by 23%.

4 Conclusions

In this paper, we presented an innovative camera switching method which is based on deep learning and a temporal tracking scheme to automatically pick the most important view to be broadcasted. We showed that our YOLOv4 model, implemented for the case of ice hockey, accurately detects players, puck, net, goalie, and referees, outperforming the state-of-the-art object recognition method proposed for ice hockey by 21.57% for the puck class, with much faster inference time, achieving real-time performance. We introduce a unique camera switching scheme that weights the importance of objects in a scheme and employs temporal tracking to assign scores to the views. Our evaluations have shown that our method achieved the switching accuracy of 98% a 23% improvement over the state-of-the-art.

Acknowledgments. This work was supported in part by the Natural Sciences and Engineering Research Council of Canada (NSERC – PG 11R12450), and TELUS (PG 11R10321).

References

1. Owens, J.: Television Sports Production. CRC Press (2015)
2. Pan, Y., et al.: Smart Director: An Event-Driven Directing System for Live Broadcasting (2021). https://arxiv.org/abs/2201.04024
3. Chen, C., Wang, O., Heinzle, S., Carr, P., Smolic, A., Gross, M.: Computational sports broadcasting: automated director assistance for live sports. In: Proceedings of the IEEE International Conference on Multimedia and Expo (ICME), San Jose, pp. 1–6 (2013)
4. Wu, L.: Multi-view hockey tracking with trajectory smoothing and camera selection. Thesis, University of British Columbia (2008). https://open.library.ubc.ca/collections/ubctheses/24/items/1.0051270
5. Tohidypour, H.R., et al.: A deep learning based approach for camera switching in amateur ice hockey game broadcasting. In: 5th International Conference on Signal Processing and Information Communications (ICSPIC 2022), Paris (2022) (Accepted)
6. Ren, S., He, K., Girshick, R., Sun, J.: Faster r-cnn: Towards real-time object detection with region proposal networks. In: Advances in Neural Information Processing Systems, pp. 91–99 (2015)
7. Bochkovskiy, A., Wang, C., Liao, H.M.: Yolov4: Optimal speed and accuracy of object detection. In: Proceedings of the IEEE Computer Society Conference on Computer Vision and Pattern Recognition (2020)
8. 2019 IIHF Ice Hockey World Championship, IHF Worlds 2021, https://www.youtube.com/c/IIHFWorlds/videos. Accessed 5 Feb. 2022
9. Redmon, J., Divvala, S., Girshick, R., Farhadi, A.: You only look once: Unified, real-time object detection. In: Proceedings of the IEEE Conference on Computer Vision and Pattern Recognition (CVPR), pp. 779–788 (2016)
10. YOLOv4 darknet: GitHub repository. https://github.com/AlexeyAB/darknet
11. Digital Research Alliance of Canada (the Alliance) state-of-the-art advanced research computing network. https://alliancecan.ca/en. Accessed 29 Sept. 2022
12. Faster-RCNN. GitHub repository. https://github.com/jwyang/faster-rcnn.pytorch.git. Accessed 29 Sept. 2022

Phisherman: Phishing Link Scanner

Christian Angelo A. Escoses[2], Mark Renzo R. Magno[2], Hannah Mae P. Balba[2],
Neil C. Enriquez[2], and Marlon A. Diloy[1,2(✉)] (iD)

[1] National University, Manila, Philippines
madiloy@national-u.edu.ph
[2] NU Laguna, Calamba, Philippines
{escosesca,magnomr,balbahp,enriqueznc}@students.nu-laguna.edu.ph

Abstract. Phishing intention is to track and steal sensitive information and people often use the internet to purchase online. As a result of the increased demand for digital security and the growing number of phishing scams, the researchers decided to create a phishing link scanner. Current phishing prevention technology was studied, and with gaps identified, machine learning algorithms were used to improve and make it available to the public. The researchers created a phishing link-scanner application that runs in the background and alerts users to potentially malicious links. Implementing machine learning allows the application to adapt to uniquely structured emails and messages. Phisherman was validated using the ISO 9126 software quality model to test the application's quality, as part of the agile development methodology. The application was rated Highly Acceptable in terms of functionality, reliability, usability, efficiency, maintainability, and portability by 377 respondents, with a weighted mean of 4.5. The project met its goal of effectively verifying phishing links and informing users.

Keywords: Phishing Link Scanner · Machine Learning · Phishing · Desktop Application

1 Introduction

With the digital age expanding quickly and offering a wide range of services and people can meet all their demands online. The demand for transmitting sensitive and important data increases as internet usage grows quickly. This has made it prone to security attacks. To relate this, phishing is popular in social engineering techniques that are used by phishers to acquire sensitive and important data. Phishing is mostly from email, websites, and software [11]. In email-based phishing, attackers send millions of emails to millions of individuals to deceive at least thousands of them and gain information [11]. Most emails claim to be reliable sources. Users are persuaded into visiting phishing websites by the links in phishing emails. In website-based phishing, a legitimate website is copied to trick people into disclosing personal data. Through various social networking sites, including Facebook or Twitter, users can access phishing sites.

The increase of phishing attacks has grown over the past few years, with the phishing attacks hitting an all-time high in 2021 with APWG seeing 316,747 attacks, this was

É. Renault and P. Mühlethaler (Eds.): MLN 2022, LNCS 13767, pp. 153–168, 2023.
https://doi.org/10.1007/978-3-031-36183-8_11

the highest in recorded history [10]. During the pandemic, people were forced to stay at home because of schools, businesses, and workplaces shutting down, this made the internet become the main platform for social interaction and the main target for social engineering scams. With that, more people are being exposed to well-composed phishing emails that give them a sense of urgency to click on links [4]. The human factor is a crucial factor in phishing, even specialists make mistakes thinking their security measures can protect their average consumers [9]. Attackers often find a workaround based on common knowledge of how people view a secure network. An SSL connection can be used to trick users into thinking that their connection is secure, a simple lock icon would indicate a secured connection but can be intentionally placed to trick them to gather their information. In avoiding and mitigating the risks of phishing attacks, which try to acquire confidential information by sophisticated methods, techniques, and technologies such as phishing through content injections, online social networks, and other devices [2].

Here are some of the Famous Phishing Incidents from History: Facebook and Google, this is a major issue. A single email fraud from Lithuania cost Facebook and Google $100 million, two of the biggest internet companies in the world. Although an arrest was made, the incident demonstrates how vulnerable even the most cutting-edge technology companies are to phishing assaults [13]. A leak from Sony Pictures, a massive data breach involving over 100 Terabytes of sensitive corporate information occurred at Sony Pictures in 2014, costing the corporation well over $100 million. The top-level employees who clicked on the dangerous attachments in the phishing emails were tricked by the scammers, who purported to be their coworkers. The assault specifically made use of a bogus Apple ID verification email. The phishers discovered passwords that matched those used for the Sony network using a mix of LinkedIn data and Apple ID logins, which is a prime illustration of the significance of having unique passwords for various online accounts [12]. World Cup 2018, regarding phishing attempts during the 2018 World Cup in Russia, the Federal Trade Commission issued the following statement. To receive the prize, the victim was instructed to input their personal information after the con artist claimed they had won World Cup tickets in a lottery. A few rental frauds were also reported around the same period. During the athletic event, cybercriminals offered insanely low prices for real estate using stolen email addresses from Russian landlords. When a "lucky buyer" agreed to the offer, the details of his or her credit card were taken [3].

There are many phishing detections on the market that have been approached with deep learning algorithms that have delivered promising results. One way to address the phishing problem is implementing the DMARC or Domain-based Message Authentication, and Reporting, and this protocol is intended to protect email systems from unauthorized use. With this, we can monitor the misuse of email sending and can protect users from phishing. DMARC assists in verifying the origin of emails and preventing fake emails from being received and opened [7]. However, only a small percentage of businesses have adopted the protocol, and even fewer have implemented it correctly. As a result, phishing emails and links are still being sent and many people are still being tricked into clicking on them.

Applications such as Avanan offer anti-phishing software for cloud-hosted email, tying into your email provider using APIs to train their AI using historical email [5]. The service analyzes not just message contents, formatting, and header information, but evaluates existing relationships between senders and receivers to establish a level of trust. And Barracuda Sentinel leverages mail provider APIs to protect against phishing as well as business email compromise (BEC). Because compromised email accounts tend to lead to more phishing attempts or further account-based attacks, Barracuda's focus on minimizing further damage as a result of a successful phishing attempt has more value than relying solely on prevention. Barracuda also provides brand protection and domain fraud prevention through DMARC analysis and reporting (Table 1).

Table 1. Features Comparison of Phishing Detection tools with Phisherman Application

Features	Ironscales [2]	Avanan [1]	Barracuda Sentinel [3]	Phisherman
Email Monitoring	✓			✓
Spam Detection	✓		✓	✓
Threat Response	✓			✓
Report/Analytics	✓		✓	✓
Threat Protection	✓			✓
AI/Machine Learning		✓		✓
API		✓		✓
Activity Dashboard		✓		✓
Anti-Spam		✓		✓
Anti-Virus		✓		✓
Application Security		✓		✓
Compliance Management		✓		✓
Data Verification		✓		✓
Email Attachment Protection		✓		✓
Email Filtering		✓		✓
Encryption		✓		✓
Event Logs		✓		✓
Incident Management		✓		✓
Policy Management		✓		✓
Quarantine		✓	✓	✓
Real-Time Analytics		✓		✓
Real-Time Reporting		✓		✓

(continued)

Table 1. (*continued*)

Features	Ironscales [2]	Avanan [1]	Barracuda Sentinel [3]	Phisherman
Reporting/Analytics		✓		✓
Status Tracking		✓		✓
Threat Intelligence		✓		✓
Threat Protection		✓		✓
Threat Response		✓		✓
Visual Analytics		✓		✓
Vulnerability Scanning		✓		✓
Whitelisting/Blacklisting		✓		✓
Allow/ Block List		✓		✓
Encryption		✓		✓
Fraud Detection			✓	✓
Audit Log			✓	✓
Email Archiving & Recovery			✓	✓
Hybrid API				✓
Notify Phishing Link				✓

These tools give strong security against phishing links, but they tackle different parts of a problem. The Avanan aims for cloud-hosted email and uses AI for detection, this becomes a problem when emails that are uniquely created get bypassed, while the Barracuda Sentinel focuses more on the disaster management side. Creating an interface that provides users with the first line of defense against phishing has become a priority and combining it with automation has become the vision for Phisherman.

Since one of the main factors of successful phishing attacks is human error, the researchers thought of developing an easy-to-use and hands-free link detector application that notifies users of suspected phishing links. Phisherman will scan every link in the.

1.1 Problems Encountered/Opportunities

Approximately 4.5% of adult Filipinos, or 5 million people, claimed to have been victims of identity theft, which could have resulted in financial losses. Furthermore, 6%, or approximately 6.7 million Filipinos, believe their identity was used fraudulently to open a financial account, indicating rising awareness of such scams. With that, the information integrity of the country has taken a toll and the number is still increasing. Current technologies are ineffective against regular users; most people experiencing financial difficulties are duped into clicking links that promise them money or assets. Because of Today's evolving technology and information transparency, phishing scammers can send convincing emails and messages. With the digital trend of machine learning, the study of these email structures will be adapted to protect regular internet users on various platforms. Victims of phishing scams claim to have lost money not only for themselves but also for their company, which has an impact on the integrity of businesses and their relationships with their employees.

1.2 Research Objective

The main objective of this study is to develop a Phisherman that scans links through selected link checker APIs while running in the background. Specifically, the study aims to: (1) Develop an interface for Phisherman; (2) Utilize Hybrid Phishing Checker APIs for scanning malicious links; (3) Test the software using ISO 1926.

1.3 Scope and Limitations

The study focuses on the creation of solutions for identifying fraudulent links to address the issues with the widespread phishing scams that happen globally. This research will focus on protecting users from scams considering today's phishing attacks are more inventive and convincing nowadays. Overall, this program will help users in being more aware of potential phishing schemes, particularly for individuals who lack technological literacy.

1.4 Significance of the Study

This application will promote computer safety for people who are computer illiterate that would not distinguish a phishing link from an authentic link. Users will be given a safety measure when opening emails and prompt messages to notify them when the link is suspected to be a phishing link.

With this research, we can contribute to the nation's innovation toward cybersecurity regarding phishing attacks. This application will safeguard the Filipino people and their operational activities by preventing phishing and keeping them from losing time, and money.

Additionally, this will also prevent information leakage and maintain the integrity of our nation. Moreover, this research will use existing phishing protection frameworks and innovate them using AI with ease of use and convenience. Many existing phishing detection applications would require human intervention to scan and verify the links, while this application would do it hands-free with any window open.

2 Research Methodology

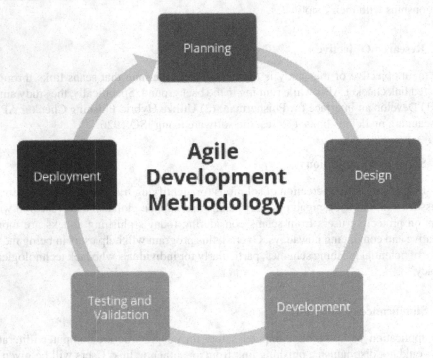

Fig. 1. Agile Development Methodology

This research will employ the Agile Software Methodology, which includes development-related attributes. The agile model also involves the iterative and incremental method, it works iteratively because of subsequent iterations and incremental in terms of sprints. This model is suitable for projects with design uncertainty. The project methodology includes properties such as:

2.1 Planning

The planning component of this research entails developing a process flow for verifying scanned URL links. The analysis of using cloud storage for the application's blacklisted and whitelisted websites is also determined. This property considers the design of the application's system architecture, which determines how each element interacts with the application's actors. The diagram below depicts the desktop application's system architecture (Fig. 2).

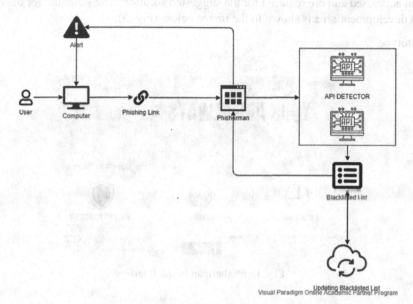

Fig. 2. Phisherman System Architecture

2.2 Design and Development

Coding and testing of the computer program are included in the development. A first iteration or working prototype will be built and tested. When the intended final output is reached, perform validation findings as shown in Fig. 1, and this first result will be continuously enhanced or improved. Specification happens when it comes to development as well. There are determined hardware requirements as well as the necessary software tools. For the researcher's part, the software and development tools to be used are Cloud SQL for the database management system required for the blacklisted or potential phishing websites or links, Qt for creating multiplatform programs that run on all major desktop platforms and graphical user interfaces (GUIs), and C++ for developing applications.

The application is specifically made for computers running the Linux and Windows operating systems to meet the system requirements of the target user. While on the hardware requirements, CPU: 800 MHz minimum, SSE2 enabled. Athlon 64, Sempron 64, Turion 64, and Phenom CPU generations from AMD, as well as the majority of contemporary Intel x86 processors, are included in this. RAM: 1 GB, 2 GB (64-bit OS) (32-bit OS), and Minimum 250 MB of free storage space. It will start on the user's computer if it complies with the hardware and software requirements.

Once all the hardware and software requirements have been established, the design and development process will start. Along with the design, the processing is the development, the Phisherman will be running into the user's computer background for the design. In the last stage of development, another interview was held. This is done to make sure that the system development methods are in line with the features that have been addressed and are required for the suggested solution. The schedule for the design and development area is shown in the image below (Fig. 3).

Prototype

Fig. 3. Phisherman Initial Interface

The initial screen of the Phisherman is where the user can see the detection history, stop the scanning, and whether to turn on or off the Real-time protection (Fig. 4).

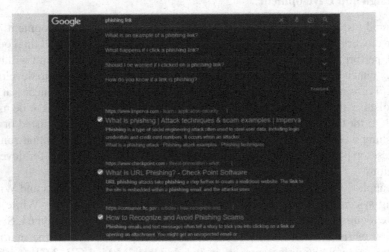

Fig. 4. Detection of trusted websites.

Once the user encountered a trusted website, there will be a green check on the left side of the link (Fig. 5).

Once the user encountered a trusted, phishing link, and unknown links they will be labeled as check, exclamation point, and question mark respectively (Fig. 6).

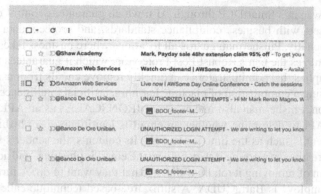

Fig. 5. Detection of trusted, phishing, and unknown email.

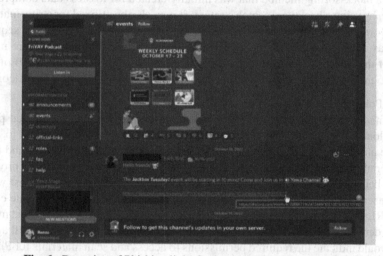

Fig. 6. Detection of Phishing links from other software or application.

Once the user hovers over the link it will show you if it is a trusted or phishing link. When it turns green it is a trusted site, otherwise, it is a phishing link or an untrusted link.

API Phishing Detection Efficiency

The email was transferred to the cloud by users. However, Office 365 or email's built-in security is not enough. Because they weren't designed for the cloud, traditional solutions like secure email gateways frequently reduce the effectiveness of built-in security. They also do not properly defend against attacks in internal emails. Deploying like an app, the Phisherman approach to securing organizations on the cloud solves these issues and more. The Phisherman process scans after the default security but before the inbox. It just means that a Phisherman-based API solution can prevent malicious emails from entering the inbox. Threats in messaging and file applications are also protected by Phisherman. By this, there is no need for a complicated configuration setup process. Phisherman also utilizes Avanan's SmartDLP through API, which guarantees control of all data in the cloud, meets with business compliance standards, and protects data from illegal access. Additionally, Phisherman guards all locations where business takes place. And, Phisherman utilizes the nan's Robust Post Protection, which it guarantees total defense in depth. It implies that administrators can quickly remove emails from users' inboxes. Along with using a combination of machine learning, artificial intelligence, and human threat detection, Phisherman also makes use of IRONSCALES as an API. Phisherman can use this API to determine whether an email is legitimate or a phishing effort by considering details such as the time it was sent, its contents, the sender's location, and more. Users are given a warning if they believe the email is phishing, and administrators have the option of removing it totally if that is what they want to do. A further API that Phisherman employs is BarraCUDA. A suffix tree-based technique called FM-index, which is based on the Burrows-Wheeler Transform, is used by BarraCUDA (BWT). BWT is a block-sorting method that was initially created for lossless data compression; however, by using a backward search strategy, it may also be utilized for string matching. The performance of this method is independent of the size of the reference sequence, and it has a low time complexity of $O(n)$ to discover an exact match, where n is the length of the query. A complete human genome may also be stored in 1.3 GB of space thanks to the reference genome's strong compression ratio of 2.5 bits per nucleotide.

Privacy Policy

Phisherman assures that users and files will be protected with the mentioned privacy in any cloud environment, from Office 365 to Gmail, Amazon Web Services to Azure. With the utilization of APIs, emails, web navigation, etc. are to be inspected and further scanned only when there is suspected malicious intent. In addition, our data is protected, and our behavior is not tracked when we are connected to the internet at least without your consent. For the privacy of message links, some malware or phishing site conceal using URLs that's why Phisherman will provide real-time scanning options that may use more in detecting threats. There will be terms and conditions that the user will agree to which would indicate the required permissions to access and scan materials for Phishing activities.

White Box Testing

Phisherman will go through testing, where developers will design test cases to undertake white box testing that aims to validate the technical side of the application. Additionally, it checks for grammatical and syntactic errors. Unit testing will be performed to identify issues and determine whether the application meets the functional requirements. The table below shows the possible questions in white box testing.

Black Box Testing

Phisherman will also undergo Black box testing. The developer used ISO 9126 which has 6 characteristics that are shown in the figure below together with its sub-categories. It aims to ensure the most accurate results since the tester doesn't know how the system works. And it will also check if all the functionality in the interface works. The table below shows the possible question to know if the applications meet the functional requirements.

Table 2. White box testing

Item	5 Strongly Agree	4 Agree	3 Neutral	2 Disagree	1 Strongly Disagree
1. The Code is readable					
2. The variables are descriptive					
3. The application is efficient in memory					
4. The Application is OS-compatible with the majority					
5. The coding is well-optimized					

The ISO 9126 standard was used as a framework for evaluating software quality that incorporates the user's perspective and the concept of quality in use. Phisherman will be tested to ensure that it meets the specifications for functionality, reliability, usability, efficiency, maintainability, and portability. In this testing, the Likert scale interpretation and weighted mean will be used to determine the level of acceptability and quantify the results. Table 2 shows the Likert scale scoring range, interpretation, and weighted mean for the aforementioned survey.

2.3 Deployment

In deployment, it is processed once the application is ready to deploy. The developers determine the problems that need to be fixed by testing methods to identify errors and things that need to be changed. Deployment of Pisherman takes place in this area. If there are concern tickets or feedback submitted on the application regarding its use or there are in a need for adding processes and features, versions of it will be released or things need to change to be better.

Population, Sample Size, and Sampling Technique

The target population of this research consists of 20000 people, of all ages that used the internet and computer devices.

The researcher used the Slovin formula in determining the appropriate sample size. From a population size of 20,000, a sample size of 377 people was chosen to represent the population with a 5% margin of error.

$$n = N/(1 + Ne2)$$

whereas:

n = no. of samples

    N = total population

Table 3. Black Box Testing

Item	5 Strongly Agree	4 Agree	3 Neutral	2 Disagree	1 Strongly Disagree
Functionality					
1. The application accomplishes the required task					
2. The application restricts unauthorized access					
Reliability					
3. The application continues to function and recovers data in					
4. The application does not have errors					
Usability					
5. The application is easily navigable					
6. The features are easily recognizable					
Efficiency					
7. The application is responsive					
8. The application efficiently uses its resources					
Maintainability					
9. The application is easy to test					
10. The application handles updates easily					
Portability					
11. The application is easy to install					
12. The application can be transferred to different environments					

Table 4. Likert Scale and Weighted Mean for Survey

Likert Scale			Weighted Mean		
Value	Range	Verbal Interpretation	Value	Range	Verbal Interpretation
5	4.21–5.00	Strongly Agree	5	4.50–5.00	Highly Acceptable
4	3.41–4.20	Agree	4	3.50–4.49	Acceptable
3	2.61–3.40	Neither agree nor disagree	3	2.50–3.49	Moderately Acceptable
2	1.81–2.60	Disagree	2	1.50–2.49	Fairly Acceptable
1	1.00–1.80	Strongly Disagree	1	1.00–1.49	Not Acceptable

e = error margin/margin of error

$n = 20000/[1+20000(.05)^2]$

$n = 377$

The method of convenience sampling was used to develop the sample of the research under the results. According to this method, convenience sampling is a non-probability sampling method that involves using respondents who are convenient to the researcher. It may be applied by stopping random people on the sidewalk or street and asking them to rate the questionnaire questions. In addition, convenient sampling is usually a low cost and easy to gather information.

3 Results and Discussion

Machine learning is used by Phisherman to detect fraudulent emails and potential phishing scams. It allows the scanned links to automatically run down different APIs, allowing for multi-feature adaptation of phishing security apps. The features include (1) a user interface for description and how to use link scanning and whitelisting, (2) detection of fraudulent emails and suspicious links, (3) link verification, and (4) post-event data protection. Tables 3 and 4 show the results of the initial testing of the prototype of the desktop application Phisherman. The tables below returned a grand mean of 4.54 for the black box testing, which is interpreted as Highly Acceptable, and 4.50 for the white box testing, which is also interpreted as Highly Acceptable, after careful evaluation of selected respondents (Tables 5 and 6).

Phisherman will be divided into two platforms, the admin side, and the users. The admins will be in charge mainly of the update and maintenance of the application. The constant updates on the blacklisted and whitelisted are observed to be necessary to keep up with the constant evolution of phishing strategies. The users that receive phishing emails will be used to further secure the list of phishing emails and with machine learning, it can automatically determine the way show phishing links are composed and where they come from.

Phisherman was first distributed to users, and the features of Phisherman were observed to be accurate in determining if the links are used for phishing scams. The application is proven to notify and inform users of phishing links that discourage them from further exploring suspicious emails. Using different APIs that scan links, emails,

Table 5. BLACK BOX TESTING SUMMARY AND RESULTS.

Item	Mean	Interpretation
1. The Code is readable	4.77	Highly Acceptable
2. The variables are descriptive	4.34	Acceptable
3. The application is efficient in memory	4.1	Acceptable
4. The Application is OS-compatible with the majority	4.66	Highly Acceptable
5. The coding is well-optimized	4.82	Highly Acceptable
Grand Mean	4.54	Highly Acceptable

Table 6. BLACKBOX TESTING SUMMARY AND RESULTS.

Item	Mean	Interpretation
Functionality		
1. The application accomplishes the required task	4.67	Highly Acceptable
2. The application restricts unauthorized access	4.32	Acceptable
Reliability		
3. The application continues to function and recovers data in	4.81	Highly Acceptable
4. The application does not have errors	4.56	Highly Acceptable
Usability		
5. The application is easily navigable	4.72	Highly Acceptable
6. The features are easily recognizable	4.35	Acceptable
Efficiency		
7. The application is responsive	4.85	Highly Acceptable
8. The application efficiently uses its resources	4.13	Acceptable
Maintainability		
9. The application is easy to test	4.62	Highly Acceptable
10. The application handles updates easily	4.42	Acceptable
Portability		
11. The application is easy to install	4.32	Acceptable
12. The application can be transferred to different environments	4.21	Acceptable
Grand Mean	4.50	Highly Acceptable

and possible scans are also proven to improve the email protection of users, the API of Barracuda Sentinel also offers Phisherman risk management if the user still clicked on the link even when there are warnings indicated.

The agile development methodology was useful in helping polish the application in development. Each scrum meeting made the team a clear goal on the application which gave each member a setting pace and each sprint went through careful validation of modules. These modules were then distributed to different users that receive phishing emails regularly for implementation.

4 Conclusion and Recommendation

4.1 Conclusion

In this paper, we have built an application called Phisherman and provided a unique heuristic method to identify phishing attacks. The application's capabilities include the ability to alert users when a malicious link is present and the utilization of a hybrid API. The two tests that may be used to determine if an application meets its functionality are also included in our explanation of our methodologies. To recognize the harmful links, there are additional whitelists and blacklists.

The main advantage of our application is that it can detect phishing sites automatically and it applies to all sites. Additionally, it employs a hybrid API, which most anti-phishing applications or methods lack.

In terms of functionality and performance, we think there are many opportunities to enhance these results. We can add additional features in the future that are more unique and beneficial to the users. Because technology is always evolving, there are many methods to make improvements. To sum up, Phisherman is very helpful in identifying phishing websites and reducing the number of people that fall victim to phishing.

4.2 Recommendation

This study, showing the efficiency of the application to the online consumers or the users, is conducted to help the user to be more cautious on every link they open or click. The results will dictate how we improve what we have done in the application of the Phisherman. This paper will help future researchers to discuss the study of phishing detection using two APIs and how they prevent human error. For future developers, it may be used as a reference to how they improve or come up with a new feature for preventing malicious links. In addition, future researchers and developers study the needs of the user and analyze the events of cybercrime, mostly the phishing attack that steals the user's information it may help them to add new security to prevent phishing attacks.

References

1. Barracuda Sentinel reviews, Slashdot. https://slashdot.org/software/p/Barracuda-Sentinel/. Accessed 20 Oct 2022
2. Catal, C., Giray, G.: Applications of deep learning for phishing detection: a systematic literature review, June 2022. https://doi.org/10.1007/s10115-022-01672-x. Accessed 4 Oct 2022
3. Healy, C.: World Cup-themed phishing attacks multiply, Corrata, 3 July 2018. https://corrata.com/world-cup-themed-phishing-attacks-multiply/. Accessed 17 Oct 2022

4. Mcclain, C., Vogels, E.A., Perrin, A., Sechopoulos, S., Rainie, L.: Pew Research Center, 1 September 2021. https://www.pewresearch.org/internet/2021/09/01/the-internet-and-the-pan demic/. Accessed 4 Oct 2022

5. Getapp.com. https://www.getapp.com/security-software/a/avanan/features/. Accessed 20 Oct 2022]

6. Getapp.com. https://www.getapp.com/security-software/a/ironscales/features/. Accessed 20 Oct 2022

7. https://swgfl.org.uk/resources/phishing-tackle/technical-phishing-prevention/

8. https://thehackernews.com/2021/09/how-does-dmarc-prevent-phishing.html#:~:text=The%20DMARC%20standard%20prevents%20this,traffic%20and%20quarantine%20suspicious%20emails

9. Jakobsson, M.: The human factor in phishing. Privacy & Security of Consumer Information (2007). Accessed 4 Oct 2022

10. Phishing Activity Trends Reports. https://apwg.org/trendsreports/. Accessed 6 July 2022

11. Rao, R.S., Ali, S.T.: PhishShield: a desktop application to detect phishing webpages through heuristic approach. Procedia Comput. Sci. **54**, 147–156 (2015). https://doi.org/10.1016/j.procs.2015.06.017. Accessed 4 Oct 2022

12. Sony hackers used phishing emails to breach company networks, Tripwire.com. https://www.tripwire.com/state-of-security/sony-hackers-used-phishing-emails-to-breach-company-networks. Accessed 17 Oct 2022

13. Huddleston, T.: How this scammer used phishing emails to steal over $100 million from Google and Facebook, CNBC, 27 March 2019. https://www.cnbc.com/2019/03/27/phishing-email-scam-stole-100-million-from-facebook-and-google.html. Accessed 17 Oct 2022

14. Check Point Software Technologies: How it works: universal threat management and security, Avanan.com. https://www.avanan.com/how-it-works. Accessed 11 Nov 2022

15. Alwari, U.K.: Email protection, Avanan.com. https://www.avanan.com/docs/create-email-pro tection-policy. Accessed 11 Nov 2022

16. Fuchs, J.: Why should I choose Avanan? Avanan.com. https://www.avanan.com/blog/why-should-i-choose-avanan. Accessed 11 Nov 2022

17. Witts, J.: IRONSCALES: a comprehensive deep dive, Expert Insights, 25 January 2021. https://expertinsights.com/insights/ironscales-overview/. Accessed 11 Nov 2022

18. Klus, P., et al.: BarraCUDA - a fast short read sequence aligner using graphics processing units. BMC Res. Notes **5**(1), 27 (2012)

Leader-Assisted Client Selection for Federated Learning in IoT via the Cooperation of Nearby Devices

Mohamed Aiche[1] , Samir Ouchani[2(✉)] , and Hafida Bouarfa[1]

[1] LRDSI Laboratory, Faculty of Science, University Blida 1, Soumaa, BP 270, Blida, Algeria
[2] INEACT Research Laboratory, Ecole d'Ingénieur CESI Aix-en-Provence, Aix-en-Provence, France
souchani@cesi.fr

Abstract. Federated learning is a form of distributed learning in which each participating node is handled as a client and is responsible for the training of a model using only its local data. The participation of inappropriate clients, on the other hand, may have a substantial impact on reliability, which would entail a significant consumption of resources. Due to this difficulty, the aspect of client selection is challenging. Most of the proposed approaches are based on selecting clients from their resource's information by gathering them in a centralized manner, which may have many drawbacks. In this paper, we propose a decentralized client selection. The gathering of resource information is accomplished through the collaboration of neighboring nodes, coordinated by the leader. The leader is elected using a leader election algorithm. Based on the gathered information, the leader then trains a lightweight deep learning model to select clients throughout the IoT context. The proposed approach has been developed and validated by experimenting with different complex scenarios.

Keywords: Federated Learning · Client selection · Leader election · Lightweight Deep Learning · IoT

1 Introduction

In this era, the Internet of Things (IoT) is a field that has truly earned its place among other technologies that have revolutionized our lives, especially, researchers' interests and studies in this field have accelerated its development in recent years. The use of IoT infrastructure promises a massive increase in the amount of generated data. According to a study conducted by the International Data Corporation (IDC), that by 2025, more than 70 ZB of data will be generated by numerous IoT devices [1]. However, utilizing this massive amount of data can give rise to an intelligent IoT infrastructure by using Machine Learning. The latter can improve the technology's efficiency and effectiveness. Typically, all

E. Renault and P. Mühlethaler (Eds.): MLN 2022, LNCS 13767, pp. 169–177, 2023.
https://doi.org/10.1007/978-3-031-36183-8_12

the data are stored in the cloud [2], which uses them to train machine learning models that will be used to make intelligent decisions over all the infrastructure. However, among the generated data there may be sensitive data whose leakage can compromise the confidentiality of users. Also, sending data to the cloud can significantly increase communication costs.

These drawbacks prompted Google to develop a new approach known as Federated Learning [3]. In this approach, each node trains its own model locally without sending data to the cloud, thereby ensuring confidentiality and lowering communication costs. Federated Learning selects in each round a percentage of clients to train their models until achieving the desired accuracy. But selecting the wrong clients can decrease the convergence speed and increase the number of rounds, which consumes node's resources. Therefore, in order to avoid this risk and improve the process of client selection, several solutions were proposed. The majority focused on ideas that are based on gathering information about the resources of the nodes by the server. The latter then uses them to select clients according to this criterion. However, if every client begins sending information to the server and relies on it, the situation may suffer from bottlenecks, latency, and a single point of failure.

With the disadvantages listed above, our approach becomes necessary. The basic idea behind it is to decentralize the client selection process by assigning a leader to each network. The task of gathering information about the resources will be managed by the leader. Thus, the leader then uses a lightweight Deep Learning algorithm to determine the eligibility of nodes.

The structure of this paper is as follows. In Sect. 2, we present various existing works on client selection, as well as decentralized approaches that have been carried out in federated learning. Their benefits and drawbacks are also discussed. Section 3 illustrates the general architecture of our approach and details its steps. Finally, simulations and experimental results are presented in Sect. 4, followed by conclusions in Sect. 5.

2   Related Work

Federated learning was the result of research conducted in 2016 by McMahan et al. [3]. The aim was to eliminate the drawbacks of traditional approaches regarding privacy and communication costs. However, this approach ran into a number of issues that could affect the quality of learning. Several works, such as [4–6], focused on various aspects, such as communication, privacy, and statistical challenges. But even so, the limitation of their work on common aspects pushed Wahab et al. [1] to move to new ones. The importance of client selection was demonstrated in their studies. According to [7], client selection is a crucial phase that can have a big impact on how well the learning goes when choosing clients with low resources or a bad reputation. The high impact of this aspect and the difference it can make have motivated us to focus our research in this aspect.

Client selection was first introduced in [3]. It is known in the literature as VanilaFL. The idea was to randomly select a number of clients. Nevertheless, the latter has the downside of being luck-based. Because there is a high probability of having poor clients that damage the training. In fact, in IoT, most clients have limited capacities and can become out of power at any moment [1]. Several criteria can be the reason of the approach's failure. Nishio et al. [8] discuss how VanilaFL can render learning ineffective due to the selected clients' poor communication and their low computational resources. According to the authors, those criteria can slow down the model uploading and downloading time. Based on those criteria, authors have proposed a new protocol that handles such challenges. The protocol focuses on selecting the maximum number of clients that can finish the federated learning steps before a deadline set by the protocol based on the clients' resource information. However, the drawback of this protocol is that it only takes into consideration whether the client can really finish before the deadline. But, despite the speed of the client, it can be vulnerable to failures. Also, it may not have the necessary criteria for the reliability of its model.

AbdulRahman et al. [9] introduced a new protocol called FedMCCS to overcome the disadvantages of the previous approach. In FedMCCS, a number of factors, including CPU, RAM, and energy, are taken into account to establish a client's eligibility for the learning process. The data quality criterion in the clients was not considered by any of the aforementioned approaches, which was mentioned by Wang et al. [10]. The model's accuracy can be significantly affected by the quality of the data, which was thought to be a highly significant selection criterion. The authors have then proposed a selection and resource allocation strategy based on the datasets' content. On the same criteria, Taïk et al. [11] propose an approach that selects and gives priority to data-rich clients. The method also evaluates the clients' contributions in each round. Each of these methods requires the resource information to be sent to the server, and the client selection is always centralized. Unfortunately, employing this methodology comes with serious risks, such as bottlenecks and single point of failure.

The problem of a single point of failure has been mentioned by [12], where the server can fail and therefore the process stops until the server returns. In another point, it is difficult to have accurate resource information before the start of learning. The change of resources is dynamic and can depend on the tasks of calculations [13]. The work in [14] was an inspiration. The authors proposed a decentralized approach by putting a fog node in each area as a local aggregator. The aggregators are in charge of the models' aggregation of the clients that are only in their area. Then pick a fog node that will be the global aggregator. This approach eliminates the drawbacks of the centralized approach and reduces energy consumption and network latency.

3   Leader-Assisted Client Selection Approach

In this paper, a new method of client selection in federated learning is proposed. It is based on decentralizing the selection step and eliminating server-client dependence. Our approach also avoids the drawbacks of the centralized approach, such as bottlenecks, latency, and a single point of failure. The method will also focus on lowering the demand on the server and leaving it to focus only on greedy computations.

The approach is inspired by the family structure and consists of allowing the clients in each group to become more familiar with each other. The devices then work together to decide for themselves which clients in their group will be eligible to participate in the federated learning process. Nevertheless, for successful decentralization and cooperation between nodes, a leader must be elected for each network. The leader will be responsible for synchronizing the task between the nodes.

The election of a leader is done using a leader election algorithm. The latter also coordinates their assistance to gather information from the nodes. Each free node, or the one with the least use of its computational resources, sends a request to the leader to provide his assistance. The leader then randomly selects some clients for which this assistant is responsible and finally broadcasts his address to them. The clients will then begin to send information to the assistant about their activities, resources, crashes, and communication frequency, set for each period. After the deadline, the leader requests the assistants to return the gathered information to be proceeded to the final stage. Next, the leader will apply a lightweight machine learning algorithm and carry out its selection. At each round, the leader chooses the best assistant in terms of collections to be the new leader for the next round.

To summarize, the approach is decomposed into several steps, which start first with the leader election, then the assignment of the assistants, followed by gathering resources information about the nodes, and finally the selection by depending on these resources to select our clients. Figure 1 illustrates a brief architecture of the proposed approach, where we consider the following assumptions.

- Our working environment contains no mobility, and the location of the clients is fixed.
- Each node in the network must have an identifier, which must be accessible to all other nodes in the network (Fig. 1).

The leader, who will coordinate our tasks, is represented by the green circle in this figure. The assistants, who will support the leader in acquiring resources, are represented by the red circle. During the data gathering process, many nodes provide their information to the assistant. Before voting on a new leader for the following cycle, the received information is sent to the leader, who processes the selection. The machine learning is then used to exploit the large amount of gathered information about resources to determine the appropriate clients.

In the following, we detail each step of our proposed approach.

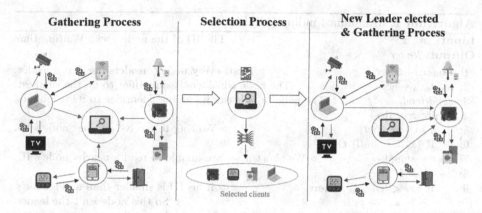

Fig. 1. The architecture of our new proposed approach

3.1   Leader Election

The first part of our approach is to elect a leader in each network. The leader is an important part of our approach. It is a node that leads the task of gathering resources by coordinating between the assistants and the other nodes to ensure the smooth running of the process. It is also responsible for gathering all the collected data and using it to select our clients. Only one leader must be elected for each network.

Several algorithms for choosing the optimal leader have been proposed, but to simplify our idea, and to just show the basic concept of our approach, we will use the Minimum Finding algorithm [15,16]. The goal is to elect the node with the smallest identifier, where each network node will at a first step considering himself as a leader and attributes its identifier to a variable *lead_min*, then each node broadcasts the value of this variable throughout the network and expects to receive other values from its network neighborhood. If a received value is smaller than the local one, the node in concern updates its variable *lead_min* with the new value, and broadcasts it again throughout the network [17]. The process concludes when a single node does not receive a value *xmin* smaller than its local one. Thus, it will be considered the leader of this network. Therefore, with this concept, each network will have its own leader. The pseudocode of the Minimum Finding algorithm is given in Algorithm 1.

3.2   Gathering Process

In our approach, every node in the network can help gather information under the leader's delegation without depending on the server. These nodes are referred to as helpers or assistants, and each set of nodes in the network is under the responsibility of a single helper. Before the allocation of assistants, the leader is the only assistant at the start of the task. After that, each node that is idle and has a low utilization rate of its resources sends a request for assistance to

Algorithm 1. Minimum Finding (MinFind).

Input: wt , x ▷ x : The ID of the node , wt : Waiting time
Output: $leader$
1: $leader \leftarrow true$ ▷ First, every node considers itself as a leader.
2: $lead_{min} \leftarrow x$ ▷ The node considers its identifier to be the smallest.
3: $send(lead_{min}, *)$ ▷ The node broadcasts its unique identifier to its neighbors.
4: **while** $true$ **do**
5:      $x_r \leftarrow read(wt)$ ▷ Waiting to receive another node's ID.
6:      **if** $(x_r == null)$ **then**
7:          $stop()$ ▷ We abort if we are unable to read a nearby node's ID.
8:      **end if**
9:      **if** $(x_r < lead_{min})$ **then**   ▷ if his ID is smaller than a neighbor's.
10:         $leader \leftarrow false$ ▷ So this node isn't the leader.
11:         $lead_{min} \leftarrow x_r$ ▷ Update the smallest identifier's value.
12:         $send(lead_{min}, *)$ ▷ Broadcast the new small identifier to all nodes.
13:     **end if**
14: **end while**

---

the leader, who then assigns randomly a percentage of clients of $1/nA$ to this assistant, where nA is the total number of assistants, and builds an assignment matrix. Finally, the leader will broadcast the helper's address to all nodes.

Each client is asked to send to the assistant of which it is assigned the resource information at each change of one of them. These include the date time of the sending, the hardware performance, the usage rate of resources, the frequency of communication with the other nodes, and the amount of the collected data since the last sending. All the nodes continue this process until they reach a deadline $tmax$. Afterwards, all the assistants send the collected information to their leader to be assembled, and proceed to the stage of selecting the clients. At each reception of a node's information by the assistant, the latter checks its storage. If it reaches saturation, it sends to the leader a request for termination to assign its nodes to other assistants. In another case, if a node contacts the helper and finds it unreachable, it informs the leader to assign the nodes for which he was responsible to be part of other assistants. Finally, if a helper does not receive information from one of these nodes after a time $tsmax$ it decreases the reputation of the concerned client, which will decrease its selection priority during federated learning.

Using this method of collection, the system can always keep track of changes in node resources and obtain reliable information for use in future selection decisions.

3.3   Selection Using Lightweight Deep Learning

We will only provide a general viewpoint and proposition in this part because it is not the spotlight of our research.

Now that the leader has received all the resource information, we need to exploit them for the purpose of selecting our participants for the learning process.

Using deep learning, we need to predict future resource utilization rates using the data gathered on the previous utilization rate to know if the node in question can have enough resources during the learning process. However, the resource constraints of devices in the IoT context may not allow for heavy computation, and the leader may struggle to deal with this [18]. Hence, we propose the use of a lightweight deep learning algorithm, for which our nodes will easily support the deep learning computation.

Using the time notion and the sending time collected during the gathering process, we will focus on Recurrent Neural Networks to deal with a time series model. However, RNNs have some limitations that can be resolved by combining it with LSTM. Taking inspiration from [18], we will propose building a lightweight model using a shallow structure with a reduced number of hidden layers (no more than two), and a small number of neurons. The output of this model is a prediction of future usage.

3.4 Electing a New Leader

After each round, a new leader will be elected, and the assistant who gathers the biggest amount of data will become the new leader. The previous leader will then broadcast the elected assistant's address to all nodes.

4 Experiments

For this part, we used OMNET++ for our simulation. The objective is to evaluate the latency and the rate of information collected between centralized and decentralized approaches.

In this evaluation, we have not yet assessed the accuracy or resource consumption metrics. Because the decentralized collection of node states is the most important part of this research.

4.1 Simulation Setup

The setup of our simulation is as follows. We have set up 60 nodes. Every node is 20 m away from the others. Clients are 1500 km away from the server. The size of the packet that contains the information is 85 bytes, and the transfer rate of the latter is 56 kbps. Finally, the medium transmission speed is 19,700 m/s.

4.2 Results

Using these parameters, sending data from one node to its neighbor results in a 7-fold reduction in latency compared to sending it to a server. The evaluation's findings are shown in Table 1 below.

Assuming that the leader and assistants have already been determined in our simulation, and that we have only two assistants. An analysis of the rate of collection of the information has been done. In comparison to a centralized approach, the results indicate the wealth of information available through decentralization in the best cases. Figure 2 illustrates this difference.

Table 1. A latency comparison between the two approaches.

Approach	Distance	Latency
Decentralized	20 m from the client to its neighbor	12.14 ms
Centralized	1500 km km from the client to the server	88.28 ms

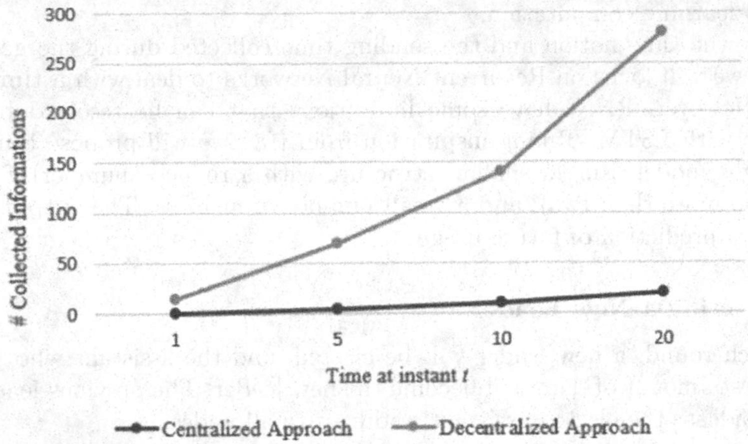

Fig. 2. A comparison of the number of the collected information in a single instant.

5   Conclusion

In this paper, we present a novel approach to decentralizing the client selection step in federated learning. Each node decides if its nearby nodes can take part in training. The aim was to avoid the drawbacks of using various server-based client selection methods.

The simulation results showed decreased latency when delivering resource information between close-by nodes as compared to other methods. It also showed how much more information could be acquired quickly using a decentralized approach than conventional methods.

In our upcoming projects, we want to refine our strategy. Selecting the node that performs the best and is nearest to all other nodes will help to start improving the leader election process. After that, we'll discuss mobility and the challenges it presents for our approach. Additionally, in order to turn the obtained data into a decision, we plan to provide a novel architecture for a lightweight deep learning model. Last but not least, we wish to utilize the leader node in additional federated learning phases.

References

1. Wahab, O.A., Mourad, A., Otrok, H., Taleb, T.: Federated machine learning: survey, multi-level classification, desirable criteria and future directions in communication and networking systems. IEEE Commun. Surv. Tutorials **23**(2), 1342–1397 (2021). https://doi.org/10.1109/COMST.2021.3058573

2. Karimipour, H., Derakhshan, F. (eds.): AI-Enabled Threat Detection and Security Analysis for Industrial IoT. Springer, Cham (2021). https://doi.org/10.1007/978-3-030-76613-9

3. McMahan, B., Moore, E., Ramage, D., Hampson, S., Arcas, B.A.Y.: Communication-efficient learning of deep networks from decentralized data. In: Singh, A., Zhu, J. (eds.) Proceedings of the 20th International Conference on Artificial Intelligence and Statistics, Proceedings of Machine Learning Research, vol. 54, pp. 1273–1282. PMLR (2017). https://proceedings.mlr.press/v54/mcmahan17a.html

4. Kairouz, P., et al.: Advances and open problems in federated learning. Foundations Trends® Mach. Learn. **14**(1–2), 1–210 (2021)

5. Li, Q., et al.: A survey on federated learning systems: vision, hype and reality for data privacy and protection. IEEE Trans. Knowl. Data Eng. (2021)

6. Li, T., Sahu, A.K., Talwalkar, A., Smith, V.: Federated learning: challenges, methods, and future directions. IEEE Signal Process. Mag. **37**(3), 50–60 (2020)

7. Du, Z., Wu, C., Yoshinaga, T., Yau, K.L.A., Ji, Y., Li, J.: Federated learning for vehicular internet of things: recent advances and open issues. IEEE Open J. Comput. Soc. **1**, 45–61 (2020)

8. Nishio, T., Yonetani, R.: Client selection for federated learning with heterogeneous resources in mobile edge. In: ICC 2019–2019 IEEE international conference on communications (ICC), pp. 1–7. IEEE (2019)

9. AbdulRahman, S., Tout, H., Mourad, A., Talhi, C.: FedMCCS: multicriteria client selection model for optimal IoT federated learning. IEEE Internet Things J. **8**(6), 4723–4735 (2020)

10. Wang, S., Liu, F., Xia, H.: Content-based vehicle selection and resource allocation for federated learning in IoV. In: 2021 IEEE Wireless Communications and Networking Conference Workshops (WCNCW), pp. 1–7. IEEE (2021)

11. Taïk, A., Moudoud, H., Cherkaoui, S.: Data-quality based scheduling for federated edge learning. In: 2021 IEEE 46th Conference on Local Computer Networks (LCN), pp. 17–23. IEEE (2021)

12. Zhao, Y., et al.: Privacy-preserving blockchain-based federated learning for IoT devices. IEEE Internet Things J. **8**(3), 1817–1829 (2020)

13. Yoshida, N., Nishio, T., Morikura, M., Yamamoto, K.: MAB-based client selection for federated learning with uncertain resources in mobile networks. In: 2020 IEEE Globecom Workshops, GC Wkshps, pp. 1–6. IEEE (2020)

14. Saha, R., Misra, S., Deb, P.K.: FogFL: fog-assisted federated learning for resource-constrained IoT devices. IEEE Internet Things J. **8**(10), 8456–8463 (2020)

15. Rahman, M.U.: Leader election in the internet of things: challenges and opportunities. arXiv preprint arXiv:1911.00759 (2019)

16. Méndez, M., Tinetti, F.G., Duran, A.M., Obon, D.A., Bartolome, N.G.: Distributed algorithms on IoT devices: bully leader election. In: 2017 International Conference on Computational Science and Computational Intelligence (CSCI), pp. 1351–1355. IEEE (2017)

17. Kadjouh, N., et al.: A dominating tree based leader election algorithm for smart cities IoT infrastructure. Mob. Netw. Appl., 1–14 (2020)

18. Agarwal, P., Alam, M.: A lightweight deep learning model for human activity recognition on edge devices. Procedia Comput. Sci. **167**, 2364–2373 (2020)

Author Index

É. Renault and P. Mühlethaler (Eds.): MLN 2022, LNCS 13767, pp. 179–180, 2023.
https://doi.org/10.1007/978-3-031-36183-8

Printed in the United States
by Baker & Taylor Publisher Services

Printed in the United States
by Baker & Taylor Publisher Services